MW01470421

Advance Praise for
BEHIND THE CLICK

"*Behind The Click* masterfully unpacks the hidden psychological motives that drive consumer behavior. Jon seamlessly connects theory to practice with prescriptive strategies that can be implemented by companies. This book is an invaluable asset for executives aiming to stay ahead in an ever-changing digital environment."

– **NIRO SIVANATHAN**, Professor of Organisational Behaviour at London Business School

"In an era overwhelmed by digital noise, Jon gives us an essential guide to not just capture, but also retain customer attention. *Behind The Click* expertly combines deep insights into digital journey optimization with practical applications of psychological principles, offering a comprehensive strategy for creating engaging customer experiences. Jon's approach is both insightful and actionable, making it an invaluable resource for marketers, digital strategists, and growth teams seeking to attract and engage the right customers in today's competitive digital landscape."

—**HEIDI DEAN**, Product Growth Manager at Adobe

"*Behind The Click* offers a compelling dive into the ever-evolving world of digital marketing, reminding both seasoned professionals and newcomers alike that success lies not just in tactics, but in understanding the underlying psychology. Jon adeptly guides readers through the intricacies of leveraging cognitive biases to craft customer-centric experiences that resonate deeply. This book is not just a manual; it's a paradigm shift, illuminating strategies that not only optimize digital journeys but transform the way we approach marketing as a whole. A must-read for anyone looking to thrive in today's dynamic landscape."

— **ISRAEL SERNA**, Channel Marketing at Autodesk

"This book is a must-read for digital leaders, particularly ecommerce and product marketing teams. While there are plenty of resources out there on shopper behavior, *Behind The Click* goes beyond theory to take those abstract ideas and offer clear, tangible tactics for optimizing digital experiences."

— **KATELYN BOURGOIN**, CEO at Customer Camp

"*Behind The Click* serves as a masterclass in consumer psychology, helping us decode the minds of our customers and leverage their complex psychological patterns. This book perfectly bridges the gap between theory and action, offering a step by step guide for readers to design digital experiences that convert!"

– **DR. KIMBERLY MOFFIT**, Relationship & Psychology Expert, Founder, KMA Therapy

"If you're looking to step up your digital game, *Behind The Click* is like a treasure map that guides you to where you want to go, blending expert knowledge with practical steps in a way that's both enlightening and incredibly approachable."

— **CHASE DIMOND**, Structured

"Jon's best work yet is a masterful exploration into the intricate psychological forces that influences online behavior, giving marketers a roadmap to creating digital experiences that are not only engaging but truly memorable. His insightful strategies for harnessing these hidden dynamics are both innovative and practical, making *Behind The Click* an essential resource for anyone looking to excel in the digital marketing space."

— **JASON WONG**, Founder at Doe Lashes, Saucy and Pughaus

"*Behind The Click* skillfully demystifies the psychological principles underlying consumer actions and translates them into practical, actionable strategies for digital success. It goes well past traditional marketing wisdom, making it essential reading for forward-thinking leaders."

— **JORDAN GAL**, Founder at Rally

"Jon goes beyond tactics and offers a repeatable framework to help you understand what drives people from awareness to purchase, and guide your audience through their customer journey. A great read for revenue-driven marketers."

— **AMANDA NATIVIDAD**, VP Marketing at SparkToro

"*Behind The Click* is essentially a friendly masterclass in understanding the heart and soul of consumer behavior online. This book bridges the gap between heady theories and real-world tactics with an engaging tone that makes you feel empowered to truly understand your optimizations, not just implement them."

— **NIK SHARMA**, Founder of Sharma Brands

"Jon's latest book really helps readers get a deeper understanding of what motivates customers, and develop the right strategies to help them to optimize the digital customer journey. *Behind The Click* is a critical read for any marketer who wants to understand the psychological forces that shape consumer decisions and succeed in today's increasingly complex marketplace."

— **ERIK HERMANSON**, Global Head of Marketing Tech & Operations at Giant Bicycle

BEHIND THE CLICK

How to Use the Hidden Psychological
Forces That Shape Online Behavior
to Craft Digital Journeys That
Delight, Engage, and Convert

R. JON MacDONALD

PRINTED WORDS

COPYRIGHT © 2024 R. JON MACDONALD
All rights reserved.

BEHIND THE CLICK
How to Use the Hidden Psychological Forces That Shape Online Behavior to Craft Digital Journeys That Delight, Engage, and Convert
First Edition

ISBN 979-8-9899356-0-4 *Hardcover*
      979-8-9899356-1-1 *Paperback*
      979-8-9899356-2-8 *Ebook*
      979-8-9899356-3-5 *Audiobook*

Here's to removing all the
bad digital experiences,
until only the good remain.

CONTENTS

Acknowledgements xiii

INTRODUCTION XV

   Who Is The Good? xviii
   Why I Wrote This Book xx
   How to Use This Book xxii
   How Not to Use This Book xxiv

1. SHORTCUTS 1

   Why Should Companies Work with Heuristics? 6
   How Can You Work within These Heuristics? 11
   At a Glance 12

2. DISCOVERY 15

   Question: Does This Company Understand My Problem? 16
   Solution: Carefully Craft Your First Impression 17

   Question: Does This Company Have a Solution to My Problem? 46

Solution: Conduct User Testing   49

At a Glance   54

3. INFORMATION-GATHERING   59

Question: Can I Find What I'm Looking For?   62
Solution: Create a Clear and Customizable Browsing Experience   63

Question: What Do I Need to Know about This Purchase?   110
Solution 1: Leave Nothing Up to Interpretation   112
Solution 2: Say It Again…and Again…and Again   128

Question: Who Else Has Already Bought This, and If I Join Them, What Does That Say about Me?   152
Solution: Show Them Exactly What This Purchasing Choice Says about Them   153

At a Glance   166

4. DECISION-MAKING AND CONVERSIONS   173

Question: Am I Ready to Make a Purchase?   176
Solution 1: Get Customers Ready to Convert with an Emotional Appeal   177
Solution 2: Hit Them with a Meaningful and Well-Timed Call to Action   186
Solution 3: Make Customers Feel Like They Already Own It through Customization   189

Question: What Should I Choose?   194
Solution 1: Remove All Barriers That Make the Choice More Complicated   195

Solution 2: Compare Wisely   206
Solution 3: Offer Simplified Options   216

Question: What's in It for Me?   223
Solution: Give Something Extra instead of Taking Something Away   227

Question: If I'm Wrong about This Choice, What Is the Worst Possible Outcome I Could Experience?   239
Solution: Build Trust and then Guarantee It   242

Question: How Easy Is It for Me to Purchase?   252
Solution: Streamline the Checkout Process   254

At a Glance   262

5. POST-PURCHASE   269

Question: Were My Expectations Met?   272
Solution: Proactive Communication for the Good *and* the Bad   273

Question: Did This Product or Service Work as It Should?   278
Solution: Create a Clear and Easy Onboarding Process   279

Question: Would I Make This Decision Again or Recommend that a Friend Do the Same?   288
Solution: Leave Them on a Good Note with Post-Purchase emails   289

At a Glance   295

Conclusion   299
About the Author   307

ACKNOWLEDGEMENTS

This book would not have been possible without the many wonderful, brilliant team members who have been part of The Good's mission over the last decade-plus to remove all the bad online experiences until only the good ones remain. Because of you, the internet is a better place.

Especially worthy of recognition is Elyse Notarianni, who made major contributions to getting this book written and whose patience, guidance, and editing along the way made it all possible.

Thank you to Jason Munger for bringing the book to life with amazing illustrations, Lexi Traylor who designed all of the wonderful example graphics, AJ Hendrickson whose copyediting and proofreading ensured poor form doesn't get in the way of function, and Caroline Appert who has helped me drive this book to successful completion.

INTRODUCTION

People make 35,000 decisions every day.[1]

Yes, even you.

But if someone asked you to name every single one of those 35,000 decisions, there's a pretty decent chance that you couldn't even name 100. Or fifty.

Sometimes, they are big decisions, like deciding if you should accept that new job offer or whether you're ready to adopt a dog. A lot of the time, they are small decisions, like choosing your outfit for the day. Most of the time, though, you make decisions that you have no idea you're even making. Hitting the snooze button on your alarm clock may feel like an automatic response when in reality, it's an unconscious decision that yes, you really do need ten more minutes of sleep.

---

1   https://go.roberts.edu/leadingedge/the-great-choices-of-strategic-leaders

Each of these decisions, conscious or unconscious, comes with all sorts of factors to consider. "If I adopt the dog now, am I going to have time to take him out during the day while I'm at work? Does this shirt actually need to be ironed, or will anyone really care if it's a little bit wrinkled? If I sleep for ten more minutes, will I have enough time to take the trash out before I leave the house?" Often, these thoughts—and their answers—run through your head so quickly and so automatically that you barely pause to notice.

There are several factors that play into every decision we make throughout the day. Multiply that by 35,000, and we'd never have time to do anything else. So our brains create shortcuts to help us make those decisions as quickly, easily, and efficiently as possible.

INTRODUCTION

This is exactly what happens every time a customer begins their digital journey.

This is where psychology and sales meet. When someone opens up their web browser, tablet, phone, or whatever technology has been invented from the time this book was published that allows people to purchase online, there are so many factors at play that affect how they experience your website or app and, ultimately, if, when, and how they choose to engage.

Bombarded with information but armed with a specific mission, users' brains automatically create shortcuts to help them meet whatever purpose they came to your company to fulfill.

Any digital journey is, on the surface, independent by nature. There's no one there to greet you at the door, walk you through your options, and answer questions in real time. You can't judge a product's quality by holding it in your hand. But that's not to say these experiences don't still happen in a different way online. Your website copy can provide the same function, as can your images, calls to action, pricing page, and product descriptions. Your navigation, your filters, and your categories can all help get your customers to the same end goal—just in a different way.

While these elements seem technical on the surface, the way users interact with them stems from something deeper. That deeper motivation can be influenced by hidden factors both in and out of your company's control. If you want to turn your users into customers, it's your responsibility to account for those factors to optimize your customer's digital journey.

WHO IS THE GOOD?

Throughout this book, when I refer to "we" or "us," I'm writing about the team at The Good.

The Good is a digital experience optimization agency that I founded in 2009 with the mission of removing all the bad online experiences until only the good remain.

Our experience has shown us that a company's digital journey is constantly evolving, and no company is ever truly done optimizing. We've not only set industry best practices, but also challenged and refined them.

The digital experience is the overall journey your customer takes from the moment they first encounter your company to long after purchase. It's far from a vague industry buzzword built on commercialized best practices and guesswork—it's a pathway that you can diagnose, track, test, and optimize through qualitative research and data-backed analysis.

Digital experience optimization is both an art and a science. It is not about diagnosing what the problem might be and then throwing in a fix that should work. It's about uncovering issues that you can prove, through research and data, exist

INTRODUCTION

within your customer's unique journey and testing solutions that yield real, measurable results.

The Good has worked with some of the world's largest companies, including Adobe, Nike, Xerox, Intel, and more, to unlock engagement and grow revenue from their digital products. We've created over $100 million in additional revenue for our clients.

Through that experience, we've learned that to employ digital experience optimization well, you have to look beyond strategies focused on boosting your conversion rate. Conversion rate optimization is one part of the overall digital journey, but it's not the sole driver in revenue growth. In fact, if you just focus on improving your conversion rate, you may encourage users to finalize their purchase, but you'll overlook the customers who never got that far down your digital journey in the first place.

Digital experience optimization pulls back on the sales funnel to capture customers from their first on-site or in-product touch point. Then, it ushers users along a journey that seamlessly guides them from initial engagement to confirmed customer to loyal advocate for your company.

I'm writing this because optimization is so much larger than just conversion. If you want to create a digital journey that gives your customer the experience they really want, you need to look at the experience as a whole. More importantly, you need to understand the hidden forces at play that influence how customers navigate that experience.

WHY I WROTE THIS BOOK

In optimization, there's a tendency to jump straight into tactics. Everyone wants to know the what—*what* tactic to deploy to enhance the overall digital experience and boost revenue. But what's often missing is the why.

This isn't the first time I've written about both the what and the why of digital experience optimization. My first book, *Stop Marketing, Start Selling*, addresses why awareness strategies and traditional marketing aren't enough to convert and provides actionable steps to shift that mindset into a sales-focused digital journey.

My second book, *Opting in to Optimization*, dives into specific optimization strategies, how to implement them successfully, and why they work from a technical standpoint.

Up until this point, my team at The Good and I have been asking questions like, "*Why* should we implement a specific tactic?" That why requires monitoring and optimizations to both ensure that tactic still serves you and optimize again when it doesn't. But there's still room to dig deeper.

To create digital experiences that hold true at their essence, regardless of changing technology or user behavior, you need

INTRODUCTION

to get to the root of what drives a successful journey in the first place. That answer doesn't live in user data—it lives in human nature.

Now, I'm taking it one step further to ask, "*Why* does the underlying thought process make it effective?" and "*Why* did that tactic elicit that specific response from that specific user?"

When we started asking these questions, we found that cognitive biases need to be leveraged to optimize almost every area of your digital experience. If you want to create a customer-centric experience that's ready to convert, you need to take into account the biases that might influence customers when they are searching for a product, evaluating its suitability, and considering its potential benefits in achieving their goals.

This approach still requires continual testing and monitoring, just like any ongoing optimization process. But factoring in these psychological principles creates a significantly stronger foundation from which to build your optimizations.

The unique perspective of this book lies in its ability to translate psychological principles into practical steps that your company can implement on your website to enhance the digital journey.

> This book's primary goal is to shift the focus from blindly applying tactics to understanding the psychology behind them.

While many resources discuss these principles in isolation, there's a scarcity of content that guides digital leaders on how to effectively leverage them. My goal is to fill this gap by drawing from our extensive experience in data-driven journey optimization. Rather than relying on fictional stories and wireframes, I'll showcase genuine case studies and examples that align with the challenges and opportunities faced by professionals in e-commerce, SaaS, and other digital industries.

In addition to improving your website or app's performance, the insights gained from this process can have a broader impact on your entire company. These learnings aren't confined to just your digital experience—you can apply them both online and offline. While my primary focus is on optimizing your digital journey, keep in mind that the knowledge you gain can be leveraged across various aspects of your business, from research and development to customer service.

HOW TO USE THIS BOOK

Deploying tactics without comprehending the underlying psychology is like building a house without a solid foundation. It might seem stable for a while, but it's not built to withstand the test of time.

So my vision for this book is to equip readers with an understanding of the psychological principles that drive

INTRODUCTION

many successful digital journey optimization tactics and offer insights into why they work (even if the customers themselves are unaware of these underlying factors).

This book includes four main sections, revolving around the four main phases of a user's customer journey:

- Discovery
- Information-Gathering
- Decision-Making and Conversion
- Post-Purchase

Each chapter promotes a customer-focused approach by putting the company into the mindset of the customer. I state the questions your users might ask themselves at each stage of the journey. Then, I explore how you, as a company, can help them answer that question.

The ultimate goal is to guide the customer toward answering yes.

While many of these psychological principles and optimization strategies are presented in broad terms, this book is not meant to be a checklist for psychological digital journey optimization. Instead, I encourage you to understand the underlying principles behind consumer behavior and apply the knowledge that pertains to your specific, well-defined target audience in a way that resonates with them.

By understanding the "why" behind these strategies, you can adapt and tailor them to your specific circumstances. Then, you can continuously assess and ensure the tactics are still aligned with your objectives and are delivering the desired results on your website.

What works for one company may not work for another, and "best practices" aren't necessarily universal truths. However, the underlying psychological principles can be universally applied.

My objective isn't merely to help you boost sales. My focus, and now yours, is to help your ideal customers find the right product for them quickly and effectively, leading to increased sales in a more meaningful way.

HOW NOT TO USE THIS BOOK

Put simply, the goal is not to use psychological principles to trick customers into buying more than they came for or choosing your company over a competitor's.

It's essential to avoid using these techniques for unethical purposes, often referred to as "Black Hat" tactics. This is just as much a practical consideration as it is an ethical one. If your company were to employ these methods to manipulate users, you would:

INTRODUCTION

- Attract customers outside of your target market who would ultimately be unhappy with your product
- Increase return volume
- Plummet overall customer sentiment by increasing unsatisfied customers
- Waste time, money, and resources that could have otherwise gone into attracting those who *do* need your product or service

Conversion for conversion's sake may bring in money in the short term but will do more harm than good in the long run.

At the end, you should have a solid understanding of the psychological principles influencing your customers, as well as an arsenal of strategies to account for these biases. The goal is to guide your customers through their decision-making process toward an action that best achieves their goal—purchasing a product that fits their needs (which, incidentally, is your goal too).

To fully grasp and effectively utilize the psychological principles and optimization tactics detailed in this book, it's essential to understand the mental shortcuts our brains use to navigate the vast amount of information we encounter daily. This insight will give you a richer understanding of how your customers think and act, allowing you to implement strategies that really resonate.

SHORTCUTS

A T ANY GIVEN MOMENT, THERE ARE MILLIONS OF things happening around us that we aren't aware of.

Our brains take in roughly eleven million bits of information every second. But consciously, our minds can only handle forty to fifty bits of that information.[2] This means, at best, our brains process 0.00045 percent of what's going on around us.

Which, in numbers, doesn't make a huge impact. Realistically, there's a good chance your brain doesn't feel it's necessary to

---

2   https://www.britannica.com/science/information-theory

BEHIND THE CLICK

commit those stats to memory. Let's put this into some real-world perspective.

Imagine you are at a coffee shop. There are only three people in line when you walk in—a teenage girl rooting through her backpack to find her wallet, a man in his late twenties wearing a suit and staring down at the scuffs on his shoes, and a woman in her mid-forties who is texting on her phone.

After the first two order their coffees, the woman in front of you takes a step toward the register. She looks up from her half-written text and stares at the menu in silence for a few moments. She then asks the barista what flavors they have, if they have oat milk, and the difference between a latte and an Americano.

By this point, it takes everything in you not to roll your eyes so far into the back of your head that they get lost in there. It's annoying—she had more than enough time in line to figure out what she wanted from the menu before stepping up to the register. After she finally pays for her drink, you hurry to the counter, credit card already in hand and at the ready, and order a black coffee to go because a cappuccino, which you originally wanted, will take too long to make. The barista lets you know that they're brewing a fresh pot right now, so it's going to be a few minutes, which is fine. You wait at the end of the counter for your drink before you hustle out the door on your way to a meeting you otherwise would have been comfortably on time for had the woman in front of you figured out her order while waiting in line—just like everybody else.

All of this is what comes to mind as you apologize for being late. Looking back, you don't remember what color shirt she was wearing or what kind of phone she was texting on, even though you clearly saw both of those details while waiting in line behind her. It doesn't even occur to you to try to remember those details.

Why? Because in the context of what's going on around you, those things don't matter. The other details fell away because your brain didn't feel they were necessary to the narrative you homed in on. Even if you do later recall details that don't feel important, it's merely a faction of the information available.

That's what our brains do every second of every day—they take in everything going on in the world around us, carefully pick and choose what bits of that information to process, and use those factors to make quick decisions about what to do next. This is heuristics, and it is the core of just about everything you're going to read from here on out.

Put simply, heuristics are mental shortcuts people take, often unconsciously, to make quick and efficient decisions.[3]

---

[3] Tversky, A. and Kahneman, D. (1974). Judgment Under Uncertainty: Heuristics and Biases. *Science*. 185(4157), 1124-1131.

There are all sorts of heuristic types, many of which we will cover. But they all come back to this idea.

Instead of carefully weighing every aspect of every decision we're faced with day to day, our brains search for an easier path. The less effort it takes to process information, the easier we can make decisions and move on to the next task.

Sometimes, that means omitting information that isn't relevant to the task at hand. It may not really matter what color shirt the woman in front of you at a coffee shop picked out that morning because there could be so many other things going on that are more important and worth noticing. But in the context of what you're doing, what your goals are, and how you feel, you may miss them completely—not that you'll ever know the difference.

Other, it means drawing on past experiences to make a decision quickly. You ordered a black coffee because last time, they filled the cup and handed it to you before your payment had even been processed. The time before, you stood scrolling through Instagram while waiting for your cappuccino to hit the barista bar. So naturally, you feel like you have to wait longer for a cappuccino than for a drip coffee. The fact that the coffee was still brewing didn't fully register when you ordered. It felt like it would be faster, so it must be—even if you spent the same amount of time waiting at the end of the counter.

Here's the problem: while these shortcuts help make easy, efficient decisions, they aren't always reliable.[4] They can lead you to incorrect conclusions or push you to make the wrong decision. In the context of an online purchase, that decision is ultimately in the customer's hands. But they may not see it that way, which can lead to high return or unsubscribe volumes and negative customer sentiment for your company.

Some of these shortcuts are out of your control—you can't control the preconceived notions that your customers have about your company or product, and you can't hold their hand and make sure they're in a good mood, well fed, and properly hydrated before sitting down to make a purchasing decision. But you can eliminate any experience on your website that is going to exacerbate the psychological factors already at play.

By understanding those psychological factors at their core, you can create meaningful changes to the way your website or app functions in order to give customers the information and experience they really need.

---

4   Tversky, A. and Kahneman, D. (1974). Judgment Under Uncertainty: Heuristics and Biases. *Science*. 185(4157), 1124-1131.

WHY SHOULD COMPANIES WORK WITH HEURISTICS?

To customers, perception is often more important than reality. In other words, what a person feels is true holds far more weight than the actual truth. Companies play an integral role in shaping how consumers perceive them, and constructing a mental model can be pivotal for this very reason.

A mental model serves as a representation of your customer's digital journey in order to better understand and predict their behaviors. By ensuring their mental model aligns with their target audience, companies can craft a consistent and effective consumer experience that leads to conversion and overall customer satisfaction.

While many company sales and marketing departments understand this in theory, the execution typically leaves something to be desired.

In many companies, the first task in creating a good customer journey is often cosmetic. This leads to a time-consuming and expensive redesign that focuses on making the website *look* better with little regard to figuring out how to make it *function* better.

When you understand and work within the mental shortcuts your customers take while exploring your website, you can make decisions that shift the focus to a customer-centric experience. By actively removing any barriers that get in the way of these mental shortcuts, you ensure that the experience itself instills a subconscious level of trust.

That's where a fundamental understanding of heuristics becomes crucial. Companies walk a fine line between standing out from their competitors and recognizing web usability customs that shouldn't be violated (i.e., fitting into the norm). If you start violating these rules or heuristics, it can quickly raise red flags for users who expect the site to work a certain way and are frustrated when it doesn't.

Unfortunately, most users don't have the language to articulate precisely what makes a company's digital experience distrustful—they only know that something feels off. So they close their browser—and they probably don't come back.

When you work within these mental models and use these shortcuts to both your company's and the customer's advantage, you instill a subconscious sense of trust that's worth more than any catchy marketing slogan you can throw at the top of your homepage.

We'll break these elements down in much more detail throughout the book. But on the whole, a trustworthy digital experience does the following:

- **Feels familiar:** The internet has been around for a long time, and consumers have a general understanding of how to navigate the digital landscape, regardless of the type of website or app they are on. This comes down to the fundamental site setup.

    With e-commerce brands, for example, people expect categories in a navigation menu. They know to look to the top right corner of a site for a search bar. They know that if they need your address or phone number, they can scroll to the footer and find it.

    When those basic expectations are not met, it's an immediate red flag to consumers.

- **Does what it says it's going to do and does not introduce surprises:** If your company makes a promise to a customer, they expect you to keep it.

With pricing structures, for example, users generally understand how subscriptions work—either they pay monthly or they pay once. Introducing elements like surge pricing or unexpected fees violates those mental models and can lead to distrust very quickly.

During checkout, users expect to see the final cost of their purchase before starting the process. If a $5 shipping fee pops up after they've already input their credit card information, they're going to feel violated, like they were tricked into giving information for a situation they didn't fully understand.

- **Functions intuitively**: This is particularly relevant for SaaS products. Because there are so many products and experiences in the digital world, companies often do better fitting into the functional mold instead of trying to reinvent the wheel to be unique or different.

  For example, if you're accustomed to using Google Docs, you're trained to recognize that a small icon in the top right corner with a face means someone else is in the document or that it has been shared.[5]

---

5   https://thegood.com/insights/website-credibility/

How do you achieve this? You change the website's function to fit the experience the user already expects. You don't expect the user to relearn the journey for the sole purpose of purchasing from you (because they won't).

It's akin to the concept of a well-trodden path. Imagine you're out for a walk along a sidewalk, but directly beside you are paths worn across the grass where meandering feet made their own trail. Do you avoid those unauthorized shortcuts and stick to the main route, or are you one of those who helps create a new route?

Those crowd-created shortcuts, called "desire paths" by urban planners, hold a crucial lesson for digital marketing managers and e-commerce managers.[6] As with foot traffic, some online visitors are observant and content to follow the

---

6   https://en.wikipedia.org/wiki/Desire_path

path you've set forth. Your job is to make the signs they follow easy to see and understand.

Other online visitors are easily frustrated. They don't want to exert a whole lot of effort trying to figure out your navigation. They're in a hurry. They want more control over the customer journey. The better you can serve or sway those visitors, and the less friction involved in doing so, the more they'll convert into leads or purchase from you.

Companies that offer the best path to purchase are going to be rewarded with higher purchases.

HOW CAN YOU WORK WITHIN THESE HEURISTICS?

Consider why a company would believe they need to design an effective digital experience. Some visitors already know precisely how they want to use your site, and your task is to provide them with the tools and setup they need for a seamless experience. Others may not have a clear idea of what they're looking for. In such cases, you play a more active role in guiding their journey and helping them find what they need. It's your job to create a functional path for each.

To do that, your company needs to know your customers backward and forward. Who are they? What are they looking for? What terms do they use to find the website? Only armed with that information can you flip the focus on your website or app from one that serves the company's goals to one that serves the customer's goals.

This is where you become an advocate for the person who's not physically present in your meetings but significantly impacts your company's digital experience—the customer.

A tradition at Apple has been to always keep an empty chair in meetings to symbolize the customer's presence, even though they aren't physically there. The idea is for the team to constantly remind themselves to prioritize the customer's perspective. It's easy for e-commerce and SaaS companies to use the screen between them as an excuse to disconnect from their customers as human beings, and it's leaving a lot of money on the table. It's time to invite them back into the meeting.

This is why, from here on out, every question this book poses will be from the perspective of your customer—what do they need to know to take the next step of their digital journey? It's only by catering to those needs will you really see the conversion, company loyalty and growth your team has worked so hard to earn.

AT A GLANCE

Principles at Play

- **Heuristics:** Mental shortcuts people take, often unconsciously, to make quick and efficient decisions.

- **Mental model:** A representation of your customer's digital journey that allows you to better understand and predict their behaviors.
- **Desire paths:** A term used by urban planners to account for crowd-created shortcuts from one place to another.

In Summary

- Our brains only take in a small fraction of what is going on around us, and what they do take in is influenced by a variety of subconscious factors.
- Heuristics help us make quick, easy decisions, but they can sometimes lead to incorrect conclusions. We choose what feels correct—and that doesn't always mean it is correct.
- Understanding these heuristics and mental models can help companies create a digital experience that establishes a clear and easy-to-follow digital journey for their customers.
- Creating a digital experience that works within these mental models can instill an underlying sense of trust with your customer.
- Implementing these changes creates a customer-centric digital experience.

The discovery phase is where customers form initial impressions based on whether the company understands their problem and has a potential solution. These impressions are influenced by subconscious biases and preconceived notions, making it essential to guide users effectively.

DISCOVERY

WHEN A NEW USER FIRST FINDS YOUR COMPANY, they enter the Discovery Phase. Think of this as the moment a potential customer initially encounters your company—maybe it's a quick glance at your website, a swipe through a social media ad, or even a referral from a friend.

This stage is more than just superficial window-shopping. It lays the groundwork for how this person forms impressions, engages with the company, and ultimately decides whether to advance further down the purchasing funnel.

Let's dive into the judgments, emotional reactions, and mental notes that play out when users first engage with your digital product.

In this stage, users find themselves asking:

- Does this company understand my problem?
- Does this company have a solution to my problem?

QUESTION:
DOES THIS COMPANY UNDERSTAND MY PROBLEM?

In theory, people tend to understand that subconscious forces impact decisions, such as bias or stigma. But that's just it—*other* people are impacted. Surely we aren't.

And there's a reason we believe that while other people may be affected by subconscious forces, we aren't. It's called naïve realism.[7] At its core, it's the belief that our perception of the world reflects it precisely as it exists, untouched by personal biases or emotional influences. We instinctively feel that the way we see things is the "obvious" or "correct" view. This unfound assurance can greatly shape your customers' interactions and decisions.

---

[7] *Naïve realism : Meaning, Examples, Characteristics and Criticism.* (2019, September 10). Sociology Group. Retrieved December 2, 2021, from https://www.sociologygroup.com/naive-realism/

DISCOVERY

Consider online shopping. Customers often navigate a website thinking their view of a product, company, or service is the only clear and obvious one. This belief can cause customers to make snap judgments—for good or for bad. The information your company chooses to present, and how your company presents it, can shape this "clear" or "obvious" view the customers see.

SOLUTION:
CAREFULLY CRAFT YOUR FIRST IMPRESSION

The rules of the game are deceptively simple: You only get one chance to make a first impression. But the psychology that informs those first impressions is anything but simple. Users can determine whether a website is right for them in just half a second, making it crucial that you put your best foot forward quickly.

Because whatever impression you leave, it's going to stick.

It's not just an expression—this comes from a concept called anchoring bias, which causes people to overly rely on the first piece of information they encounter.[8] This first impression is used as a reference point or "anchor" for every

---

8   Tversky, A., & Kahneman, D. (1974). Judgment under uncertainty: Heuristics and biases. *Science*, 185, 1124-1131. https://doi.org/10.21236/ad0767426

decision to follow. When customers approach any type of digital experience, the initial information they encounter sets the tone for the entire experience.

When a visitor lands on your site, there are generally three possible outcomes:

- They feel an immediate attraction, and they throw their money at you.
- They feel indifferent, and they leave to find a more compelling option.
- They feel lost, confused, or misled, and they leave your site altogether (and if they do, 90 percent of those users will never come back).[9,10]

Every component of that initial impression will either add value to your company or take it away. Many companies falter by assuming that the first impression happens in their marketing, and that impression will carry over into the website or app. So they invest a hefty budget into honing those marketing tactics that get customers to the website or app without prioritizing the actual initial experience.

---

9 https://thegood.com/insights/importance-of-customer-experience-and-how-to-improve-cx
10 https://thegood.com/insights/ecommerce-company-value

This holds companies back for two reasons. First, not every customer will encounter your company through marketing efforts or word of mouth, so the website or app really *is* their first impression. Second, while your company's marketing may form the first impression of your brand or product, this offers a new opportunity to develop a first impression around the digital journey.

Awareness is great, but a strong digital journey is what really leads to conversions. Don't miss the opportunity to make a great first impression just because you assume product or service interest is enough to make the sale—because it isn't.

Instead, shift the focus to the customer. Here's how.

Focus on the Customer, Not the Company

One of the biggest mistakes companies make is creating websites that focus on themselves—their stories, missions, and values. While that's all well and good, it's not the first (or second, or third) thing a customer needs to see.

Customers don't often come to your website to learn more about you. Even if they do, they should be able to get all the essential information as easily and quickly as possible to move on to their real goal—solving the issue that brought them there in the first place.

Before they can even start to think about whether your company is the right solution, they need to know that you understand what they're here to fix in the first place. If a customer is looking for car parts, they want to know that you sell the pieces they need, not just used cars off the lot. Because otherwise, they're just wasting time that no one has to waste.

The more a company's website focuses on the company, the longer it takes for customers to find what they need.

Let's See This in Action

Focusing on customer needs doesn't always have to mean your need to compromise on your brand goals. Here, our team found a solution that fit both needs—leading to an almost $275k revenue increase. Let's take a look.

This company's main mission was to get customers to learn about—and ultimately purchase—their products and online program, which they feature prominently on their homepage. In theory, there shouldn't be anything wrong with that, but that placement online didn't bring the results they wanted.

When we dug into why, we found that customers weren't responding to the homepage information because many had skipped over it completely. Movement maps showed that users often went straight to the navigation bar, so they missed out

DISCOVERY

on the information about the products and program altogether. Even worse, the navigation bar frontloaded company-focused sections while burying the links that convert.

When we rearranged the navigation elements to place their online program and products to the front, where customers are drawn to first, we saw a significant increase in sign ups and product purchases.

On mobile, we took it a step further to open the product shop by default underneath a clear CTA to "Log In" or "Join the Program." Both changes ensure that users are primed to purchase as soon as they visit the website, no matter how they get there.

Digital journey optimization has little to do with rerouting a customer's online experience—it's about rerouting the experience to work best for them.

Invest in Good Design

When your website loads, customers might hesitate to trust it if it's visually unappealing or poorly designed. Even when a customer ends up there because of a raving word-of-mouth referral, they might hear about a great product but then head to a poor website and start questioning whether they're in the right place.

Cue the confusion.

Many companies assume the service can speak for itself—and while that may work for word-of-mouth, it doesn't translate to the web.

Think about the wedding industry, for example. A venue may have years of experience throwing the most beautiful weddings in town, but what if a newly engaged couple pops into their website to get more information and comes face to face with a terrible-looking website? They will think, "The venue can't even show themselves well. How can I trust them to create a beautiful wedding for me?"

In SaaS, this is even more important. The product itself is digital, which means if the interface is cluttered and confusing from the moment the customer logs on, it may impact their opinion on the service itself.

However, good design doesn't always have to mean "visually stunning." Yes, you want it to look great, but not at the expense of functionality. I meet companies every day that think they need a 9/10 or 10/10 visual website design to be successful.

DISCOVERY

In reality, you need a 9/10 or 10/10 functional design and only about a 4/10 visual design to earn the visitor's trust. Anything higher is just a bonus.

Your design doesn't need to be beautiful—it needs to be clean, trustworthy, and above all, functional.

Let's See This in Action

Any time the words "website redesign," "strategy overhaul or "user experience revamp," come out of your mouth in a business meeting, it's time to take a step back. Sometimes, you just need a refresh.

This client had no problem getting users to explore their virtual tours—the problem was they weren't going past this page to actually contact the center. Instead of completely redesigning the page, we took a step back to see what was already there. What's working to get people on the page and keep them there, and what can we add to get users to take action?

By updating the content to be more relevant and action-oriented, our team at The Good helped this client increase conversions by 25 percent.

We didn't have to start all over again. So many companies waste valuable time, energy, and money redoing their strategy over and over again every time they want a different result. Instead, they can optimize what they already have.

Design with Mobile in Mind

Mobile devices have solidified their standing as the go-to technology for online shopping—34 percent of US customers prefer shopping on mobile to any other channel (although desktop still has a strong foothold at 23 percent).[11] If people have a negative experience on mobile, they're 62 percent less likely to purchase from you in the future.[12]

For SaaS companies promoting an app, mobile optimization isn't just important—it's everything. The first tiny hint

---

[11] https://www.pwc.com/gx/en/industries/consumer-markets/consumer-insights-survey-feb-2023.html

[12] https://www.thinkwithgoogle.com/intl/en-gb/marketing-strategies/app-and-mobile/few-tips-speed-your-mobile-site-and-tools-test-it

of a clunky app experience is enough to send users back to their homepage to delete it altogether. No one has gigabytes to waste on a difficult app.

Dropbox is a perfect example of a web-based service that translates beautifully to apps. The interface is just as intuitive as its web version, if not even slightly easier to navigate because the folders and photos arrange themselves similarly to what you're accustomed to seeing in your phone's photo storage.

Plus, because it integrates into your other applications seamlessly, users can upload photos directly from their camera roll to Dropbox folders without having to transfer them to a computer first. A clunkier, less intuitive experience may have caused users to stick to the desktop-only version. Or worse—switch to a competitor with a better app.

There are a number of ways to optimize for a good first impression.

- **Load above-the-fold content first.** Organize your HTML code to load above-the-fold content first. This way, visitors will see your homepage content first—creating the anchoring impression you want—while lower-priority items continue to load in the background.

- **Make it responsive.** Responsive websites have the ability to change the layout to best fit the screen size. This makes the website more functional from any device and gives the impression that the experience is designed for each user who enters your website, no matter how they choose to access it.

- **Shorten your headlines.** Because of the limited screen real estate, headlines on your pages shouldn't exceed six words. The more words you try to cram into your headline, the less room you have for other content above the fold. Better yet, focus on high-quality visuals to draw the reader in.

Let's See This in Action

When customers buy products on their phones, it's what's above the fold that counts. You only have so long to capture a customer's attention, which means the most important information needs to fit on the screen—without scrolling.

For this client's business, only 30 percent of mobile users scrolled far enough to see the product description. The content grouping at the bottom sent the message that, by the time they reached that point, it was time to consider other options.

New customers didn't reach the product description because they didn't know to expect it there. Making the key purchase drivers more visible increased engagement and lifted conversions by 3.34 percent.

Choose Your Banner Wisely (and Once)

Rotating banners are everywhere, but that doesn't mean they work. It's a marketer's easy button—you have multiple messages, so you put them in a rotating carousel. The problem is that after the first slide, the audience no longer knows what else you're trying to say.

Research from Notre Dame University has shown that a staggeringly low 1 percent of site visitors click on non-rotating image carousels (carousels that don't change automatically).[13] Additionally, 84 percent of those clicks are on the first slide,

---

[13] https://erikrunyon.com/2013/01/carousel-interaction-stats/

meaning all the remaining slides combined receive a paltry 16 percent of the clicks. The numbers aren't much better for "fancy" carousels that rotate on their own. Auto-rotating carousels receive a bit more attention, but again, the first slide dominates, receiving 40 percent of all clicks.[14]

If you want to highlight multiple items at once as soon as customers log on, make sure it's something that prompts users to go out of the Discovery Phase and into the Information-Gathering Phase.

Let's See This in Action

At The Good, we worked with a client that wanted to direct customers to a more curated shopping experience. The goal: to create a rotating module that customers engage with, not just scroll past.

At first, our team tested creating a module around featured styles, but we found pretty low engagement. As in a 1.29 percent of recorded clicks kind of low engagement.

But even with that test running, we saw a lot of engagement in the "Category" section of the menu navigation. So we brought it forward. By taking a

---

14 https://www.nngroup.com/articles/auto-forwarding/

DISCOVERY

navigation item we already knew users wanted and bringing it to the forefront, we gave customers the search experience they needed before they even had to look for it.

Why did this work? Customers are now able to get a quick and comprehensive understanding of the company's offerings right upfront. They quickly discover what the company is and what it offers. Then, they are immediately met with a curated path to enter the Information-Gathering Phase.

From there, "Shop by Category" creates a more intentional and seemingly curated shopping experience, which makes customers feel like the website is set up for them, not like they're navigating a store in the dark. Users who know what they're looking for can find what they need more quickly and efficiently without having to think about it, and those who don't now have a clear path to explore

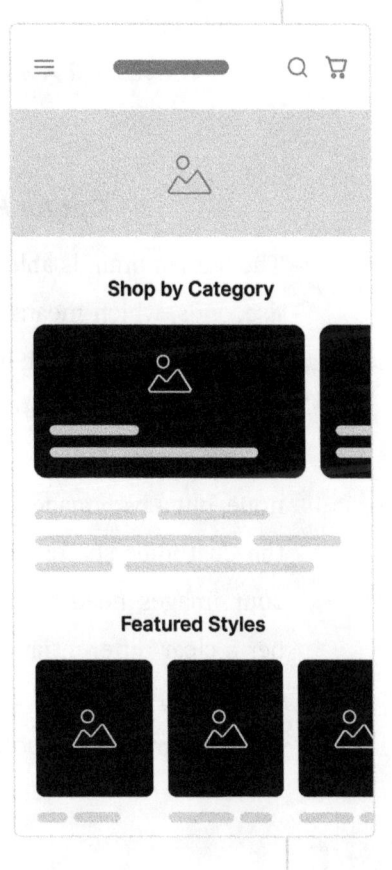

> their options. This has increased purchase confidence, add to cart, and conversions.
>
> The results? A $625,000 annual revenue boost.

Opt for High-Quality Images

The human brain is able to process images in just thirteen milliseconds, which means customers make a judgment on your company's quality with just one glance.[15,16] The quality and usefulness of those photos create an instant and lasting impression.

You might think that any website imagery is better than none, but a bad image will only distract from the goal you and the customer share: to research the product and purchase. Your images need to carry information and give the shopper a clear differentiation between one product and another. Otherwise, they will be ignored or, worse, the reason that a customer moves to a competitor's site.

> ### Let's See This in Action
>
> There's a reason great photographers don't come cheap: images influence purchasing decisions.

---

15 https://news.mit.edu/2014/in-the-blink-of-an-eye-0116
16 https://thegood.com/insights/product-image-conversions

DISCOVERY

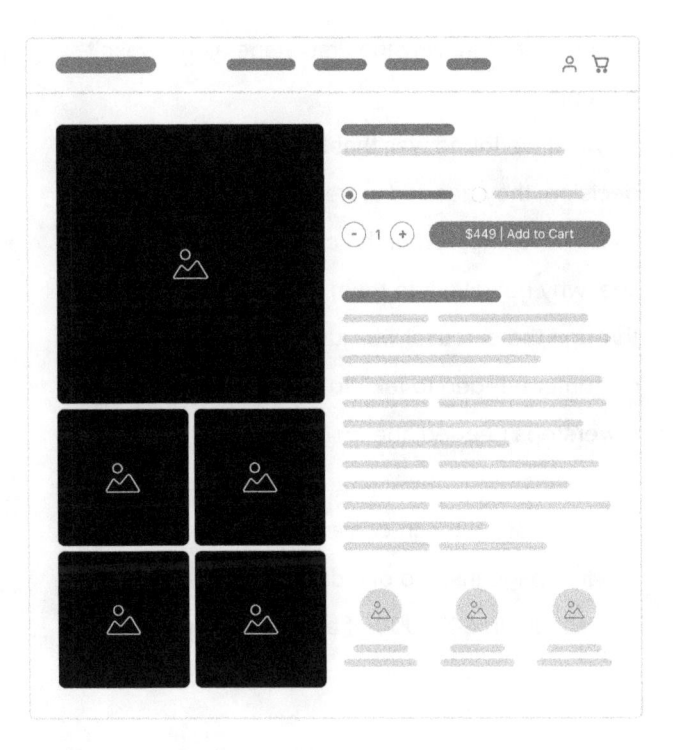

So needless to say, great images can lead to conversions. Bad images will drive someone to any competitor with a better-looking option.

For this client, we tracked movement and click maps to see what elements customers interacted with the most while gathering information and making a decision. Customers showed a high level of interest in the product carousel images.

But that was only half of the answer—images are good, but we needed to know why people were clicking through, why the first image alone wasn't enough.

What we found was that to make a purchasing decision, they needed more in-context imagery.

So if we know what the customer needs to see, why not give it to them off the bat? Clicking through the carousel was just one more step the customer needed to take to make a decision. The fewersteps needed, the more likely a customer is to convert.

By exposing the in-context images on the product page instead of hiding them in a carousel, we brought in $471,000 of annual revenue for this client.

Improve Site Speed

That half a second we talked about earlier? It doesn't start as soon as the website or site loads—it starts the moment a user moves to open it. The longer it takes to load, the less time your company has to capture their attention when they get there.

Site speed refers to the time it takes for your webpages or app to load. It's measured in seconds or fractions of seconds. Roughly 53 percent of online shoppers expect e-commerce stores to load in less than three seconds, and 79 percent of online shoppers say they won't go back to a website if they aren't happy with its loading speed.[17,18]

This doesn't just affect whether users stay on the site or in the app—it can also have a major impact on whether they end up converting. The highest conversion rates occur on pages that load in under two seconds. But in general, zero to four seconds is a good target.[19]

There are several factors that affect site speed:

- **Image size:** Images only need to be big enough to show clearly—anything more takes up unnecessary space. Compression can help you reduce the file size of your images to make your pages load quicker.

- **Number of HTTP requests:** Each element your site needed to display a page has to be requested from the website's server. More requests mean more "talking" between the server and the browser. More talking means more loading.

---

17 https://digital.com/1-in-2-visitors-abandon-a-website-that-takes-more-than-6-seconds-to-lo
18 https://neilpatel.com/blog
19 https://www.portent.com/blog/analytics/research-site-speed-hurting-everyones-revenue.htm

- **HTML and CSS code bloat:** It's remarkable how often extra code sneaks its way into your site—typically through plugins and apps, many of which you may not be using anymore and forget are there.

- **Hosting locations:** Your website consists of files stored on a host server. When you access a site, your device retrieves data from these files. The closer the server is to the user, the faster the page loads. Conversely, distant servers can lead to slow loading due to latency.[20]

All of these are within your control.

Let's See This in Action

No matter which element of the website you're optimizing at the time, site speed should always be at the back of your mind. Every image, link, metadata, function, you name it will impact site speed in some way.

Which is exactly what we saw during one client test. This client was working to optimize their homepage to increase engagement and encourage a path that

---

20   https://thegood.com/insights/site-speed-conversions/

led to purchase. We tested a few layouts to see which performed better.

And they tanked.

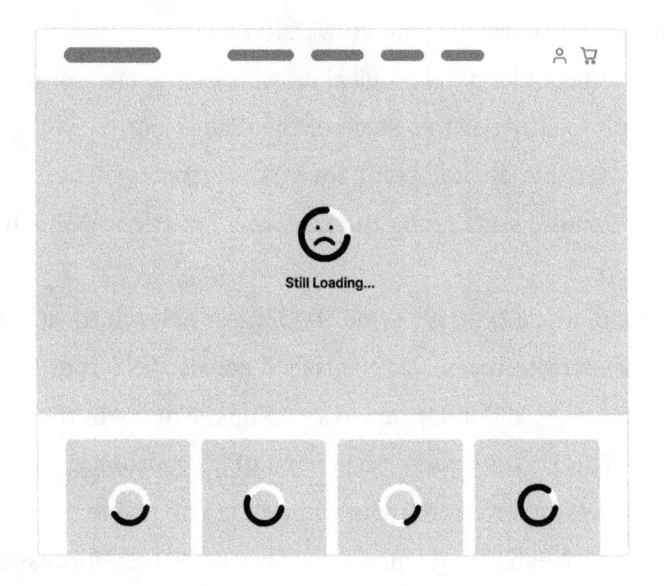

It had nothing to do with the design itself. Through the optimization process, the elements we incorporated impacted site speed significantly, which caused the test to fail altogether.

Every test brings results, regardless of whether they work or not. Let this be a lesson—you can create the world's best, most optimized website. But if it takes too long to load, very few people will stick around long enough to see it.

Don't Lead with Sales

On the surface, offering discounts seems like a perfectly fine idea. But there are many, many problems with discounting as a means of increasing your conversion rate.

In terms of its psychological drawbacks, the main problem with discounting at this stage of the digital journey is that it sets a terrible anchor point for your customers. There is a place for discounts along the customer experience—it's just not here.

When a sale hits, it's often the biggest news in town—it's on the website banner. It's splashed across the top of every product page, and it appears in a pop-up the moment the page loads. This makes sense—60 percent of US consumers say discounts are important to them while shopping online.[21]

The problem is that when a discount is the first impression your company leaves on your customers, they'll spend the rest of their digital experience thinking of you as a discount company. It devalues the products or services that *are* on sale and lowers the value of anything that *isn't* on sale.

Discounting can lower the perceived quality of your goods—a product that costs $399 looks "inferior" to a similar product priced at $600. Plus, offering promotional discounts

---

[21] https://www.statista.com/statistics/1154912/us-adults-online-shoppers-discount-how-important-coronavirus

too frequently can train your audience to wait for promotions before they buy from you.[22]

Instead of discounting your company, you want to add value. This looks different for SaaS and e-commerce companies.

- **SaaS**: Giving a taste of the experience allows customers to understand your product or service before they commit through free trials and freemium models.

    When you visit a car dealership to buy a car, the salesperson encourages you to take a test drive. They have plenty of marketing brochures and rehearsed scripts to tell you all about the vehicle, but they know that nothing sells it better than a test drive.

    Digital products can leverage the same strategy. Instead of using landing pages, demos, ads, and clever copywriting to sell the product, why not just let the user try it for free? This is one of the main tactics in product-led growth—a strategy that emphasizes the product itself as the primary driver of customer acquisition, conversion, and retention. Especially for SaaS companies, free trials and freemium models remove as many obstacles as possible to acquiring free registered users.

---

22 https://thegood.com/insights/discount-pricing

If the experience is good enough to keep them using it and the paid features are valuable enough, then the hope is that users will ultimately convert into paying customers.

- **E-commerce:** In the e-commerce realm, instead of discounting, you can offer the following:
    - **Free gift with purchase:** Getting a freebie with a purchase isn't so much a discount, but the customer is getting something for free, which sweetens the deal. This tactic is often used on first-time customers who can claim their free gift after making their first order.
    - **Free shipping if you spend over X:** This gives the impression that customers receive additional value without taking away from their evaluation of the company. Shipping options are either your best conversion tactic or the number-one reason your customers abandon their carts. Roughly 84 percent of online shoppers admit to making a purchase specifically because shipping was free and said they always add more to cart if it qualifies them for free shipping.[23]

---

23 https://www.forbes.com/sites/forbestechcouncil/2019/08/27/shipping-is-critical-to-keeping-online-shoppers-happy/?sh=52dfd4cb178c

DISCOVERY

- **Buy one, get one:** If you have excess inventory, BOGO (free or half off) is an effective way to increase customer engagement and generate sales. You can even stack the offer so that they get more when they buy more.
- **Buy one, give one:** A buy one/give one promotion is when you give something away each time a customer places an order (or for each product purchased). This promotional idea has been around for a long time, but it was popularized when shoe company Toms launched in 2006 and pledged to give away a free pair of shoes for every pair sold.
- **Free returns:** Offering free returns allows that first impression to set their mind at ease, knowing that whatever decision they make doesn't have to be permanent. Yes, it will probably increase your return rate, but it will also increase customer loyalty and set your customers up for more engaged browsing from the very beginning.

You may notice that all of these suggestions focus on adding to the value as opposed to subtracting from the price. That is exactly the point.

Let's See This in Action

Our team at The Good increased conversions by 33 percent by switching out a discount for a guarantee.

Some background: the client offered a purchase guarantee to every customer. The problem was that customers had to go to the terms of service in order to understand it.

When was the last time you read through the terms of service for, well, anything? And if users are heading there for more information, then there is a fundamental trust issue.

So we went back to the beginning of the digital experience and asked, "What are customers missing here that makes them need more information and assurance?"

We found two things:

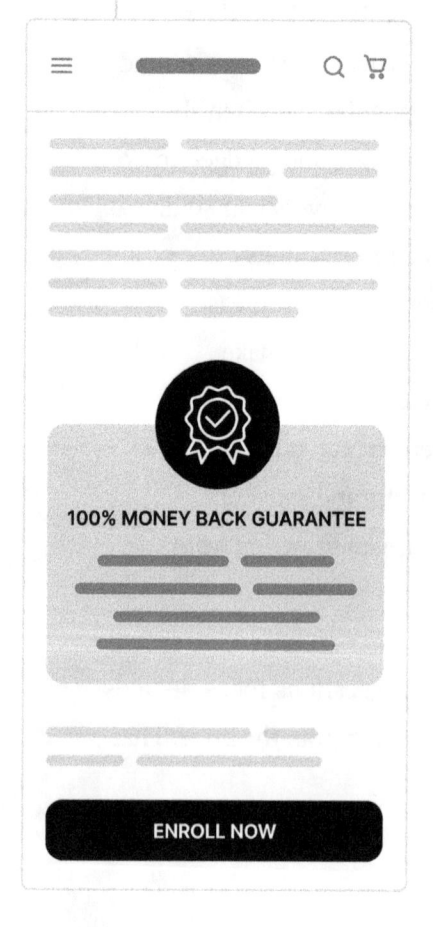

1. The content details didn't provide enough context to fully cover everything the product had to offer.
2. Even with the description, the text was so small on mobile that it was easy to miss the information.

We reworked the content to provide the assurance they needed before they knew they needed it. Instead of searching for a product guarantee, they now know they can trust it off the bat. So now, all they have to focus on is the purchase itself—which led to a 33.3 percent conversion lift.

Disable Your Pop-Ups

There's no way around it—pop-ups immediately take away value from your company, no matter how well done they are.

If you want to usher customers through the digital journey, momentum is everything. No one landed on your website or downloaded your app by accident. The user came for something specific, and just when they think they've reached it, there's a barrier to get past.

On a conscious level, it's annoying. So customers look for the exit button to return to what they came here for.

Subconsciously, customers feel a sense of "I just got here. Why are you already asking me for something?" They don't know whether they want to give their email because they aren't sure who you are yet, let alone what you do and if you can solve their problem. When you have a pop-up that says, "Enter your email for a 10% discount," for example, you:

- Immediately shift their focus to your discount, not your value
- Ask them to commit to your company by allowing you in their inbox—before they even know if they want you there
- Put a barrier into the digital experience that many customers won't have the energy to overcome

Instead, take that information and make it a part of the process. Want to gather emails? Add an email option to every page.

Think of it this way: when was the last time you walked into a store and an employee immediately stepped in front of your path and demanded you share your email? And if they did, would you be more likely to hand over your email or hightail it out of there?

That instinct to run doesn't change just because it's online.

Anyone who has seen an Instagram ad knows exactly how an unwanted pop-up feels.

You click on a link from an Instagram ad, and the first thing you're met with is a pop-up asking for your email. Before you can even dismiss it, you get hit with another popup from Instagram itself, suggesting you autofill your details. You choose not to, and just when you think you've made it to the website, there's yet another pop-up, this time about a sale or a promotion the company is running. It's easy to see how this can be frustrating for customers. People can be overwhelmed by these kinds of interruptions, especially when they come from specific sources like social media ads.

Your customers are inundated enough as it is. Depending on the how they enter your site, they may be more inundated than you think.

Keep the Copy Clear

Another point to consider is how people read online. Heat maps consistently show that readers focus on the first sentence of a large paragraph or bullet list. Then, they skim the first word or two of the rest. If it doesn't seem relevant, they skip it.

Essentially, most of the text isn't getting read. Unless it's a highly technical product where people scrutinize every detail, your most crucial information should be at the beginning.

People also assume that the first thing you tell them is the most important. So the information you really need people to know or the questions you need to answer should be front and center. Remember that first impression happens in less than a second. Make sure the copy that gets through to them is worth it.

The first copy a customer reads—and the only copy they should have to read before deciding whether you understand their problem and offer a possible solution—should answer two or three questions (and nothing more):

- Who you are
- What you do/what you offer
- Your value proposition (what makes your company worth at least learning more about)

There's no need to bombard your customers with every piece of information and value right from the beginning. Give them the information they need to stay and nothing more. Everything else can come later.

Let's See This in Action

Your company's value proposition is one of its most valuable assets—but only if you use it right. Our team at The Good used our client's value proposition to bring in more than $1.1 million in sales.

If your target audience doesn't fully understand the

value that you can provide for them, there's no reason for them to stay on your website.

The value proposition is the first thing that people on your website direct their attention toward when deciding if they need what you're selling. And if it's missing, that means you're missing out on conversion. That's exactly what was happening with this client.

So we tested a new value proposition—"the best guarantees in the industry"—and moved it to the first placement on the homepage hero carousel.

This value proposition gave customers more confidence in their buying decisions. Instead of hoping the customers know why to buy, the value proposition ensures they now know immediately—and are willing to take action because of it.

By carefully crafting your customer's first impression, you provide an anchor point from which to judge the rest of your journey.

Putting It All Together

Customers take in a number of different website elements to craft their first impression. It happens quickly, and it happens all at once. Several core factors can ensure you're adding value at first glance:

- A clear customer-centric approach
- A strong and clear initial focus point
- High-quality imagery
- Site speed
- Mobile-friendly design
- Emphasis on your company's full value
- Strong website design
- No immediate distractions
- Clear copy

QUESTION:
DOES THIS COMPANY HAVE A SOLUTION TO MY PROBLEM?

Now that the customer knows that your company understands their problem, they're ready to get into the heart of the Discovery Phase: does this company have a solution to my problem?

Many companies skip over this step in the discovery process, thinking it's assumed. At this stage, customers are not asking to see every solution you have or the specific values that set your solutions apart from the rest. That's too much information right off the bat, and it's bound to muddy the waters until they're ready.

All they want to know now is: do you have *something* that *might* work? A simple yes is all they need to move on to the next phase of the journey. The rest of the details can come later.

To answer that simple question, customers take in the information gathered during that first impression and make assumptions based on how that information fits into the preconceived notions they already had coming into this experience. This is known as the availability heuristic, which describes our tendency to use information that comes to mind quickly and easily when making decisions.[24]

Customer decisions are often influenced by recent experiences or vivid memories. For instance, a single positive experience with a company can overshadow multiple neutral or even slightly negative past interactions. Conversely, one negative experience can deter a potential customer, even if they've previously had numerous positive encounters with that company. Their preconceived notions can also have nothing to do with the company itself, instead assigning attributes—good

---

24 Tversky, Amos, and Daniel Kahneman. "Availability: A Heuristic for Judging Frequency and Probability." *Cognitive Psychology* 5, no. 2 (1973): 207–32. https://doi.org/10.1016/0010-0285(73)90033-9.

or bad—to the company because of memories involving similar products.

Understanding the assumptions that customers come into your experience with helps you craft a positive digital experience off the bat or offer an opportunity to redirect misconceptions. This benefits your company and your customer.

Why? This subconscious reliance on the availability heuristic can lead to flawed decision-making. Customers assume something must be true, whether or not it really is and whether or not those feelings came directly from your company experience or something completely separate. Not addressing these assumptions means that customers may miss out on a product that perfectly fits their needs—they just don't know it.

This is further exacerbated by confirmation bias, which is the tendency to give more weight to information that aligns with preexisting beliefs.[25]

When customers have a preconceived notion about a product or company, they naturally zero in on information that confirms these beliefs, often ignoring contrary data.

In other words, when faced with a problem, people turn to what they *assume* the solution is without necessarily looking

---

25 Healy, P. (2016, August 18). Confirmation bias: How it affects your organization and how to overcome it. Business Insights Blog. https://online.hbs.edu/blog/post/confirmation-bias-how-it-affects-your-organization-and-how-to-overcome-it

to the *real* route cause. This, in turn, makes them seek the wrong solution.

It's your job to show them clearly and early on that you have the solution they *actually* need, even if it isn't what they think they're looking for.

If you don't account for this bias from the very beginning of their journey, your customers may continue down the digital journey and choose the wrong solution because you didn't guide them well enough. Or they may leave the website completely because they assume you aren't what they need without even looking at the information provided.

That doesn't mean long paragraphs of text or lots of new information. Usually, it means simplifying, redirecting, guiding, and optimizing the information you have now.

SOLUTION:
CONDUCT USER TESTING

As a company, you have little control over what assumptions people come into your website or app experience with. But you do need to understand what those assumptions are. That's where something like user testing comes in.

Both the qualitative and quantitative data will tell you *where* people drop off your website. Then you can hypothesize *why* they drop off.

When you're observing individuals as they navigate through your website, there are valuable insights in their comments and reactions—if you know how to listen for them. There are two ways to listen for customer information: quantitatively and qualitatively.

Quantitative listening largely involves observing visitor behavior for insights: analytics events, A/B testing, website chat logs, large-scale user surveys, heat maps on FAQ pages (or all key pages, for that matter), and many more.

The unifying goal of these activities is to understand what visitors are doing/not doing and saying/not saying. Think of this "active listening data" as your bottom rungs.

To dig into *why* your visitors behave the way they do, you want to do qualitative forms of listening. These include interviewing customers or potential visitors as well as your team members who regularly interact with customers and visitors.

Participants often express their realizations and concerns during this process. They might remark, for example, they didn't realize the product had a particular feature or they had certain expectations about what the product should do that it didn't meet. Additionally, users may mention they'd heard competitors claim their products had superior durability compared to the one they're exploring, prompting them to seek more information.

These are all opportunities to see what information your pages lack and adjust accordingly.

DISCOVERY

Let's See This in Action

As it turns out, shopping for your child's baseball bat can teach you some valuable lessons in website user experience.

Our team at The Good optimized the digital journey for a company that sold baseball bats. They had no problem getting people to their website, but their conversions were lagging.

So we went to customer service.

We asked the company's customer service team what some of their most common calls sounded like, and we found out that most of the people buying from the website were parents buying bats for their kids. They had no idea what bat they needed—all they knew was that their coach said to get a bat from their company, and that was it.

Then they were surprised, confused, and overwhelmed to find hundreds of options that all looked the same. That's when they picked up the phone.

Customer service representatives would then guide them through the buying process by asking questions like, what level does your child play at? What type of hitter are they? How big is your child? What level of investment do you want to make?

If you want to find the right fit, these questions matter. A $100 bat is perfectly fine for a kid just starting, but if your kid is going for a scholarship and needs the best bat on the market, a $1,000 option may be the right investment.

Why weren't parents able to make these distinctions on their own? Well, the marketing department had come up with fancy product and feature names that sounded cool. But the parents had no idea what "BDC Bats" meant. However, you know what they could understand? The bat had anti-sting technology, so when your kid hit the ball, their hands didn't hurt.

The fact that they needed to call customer service to figure this out left customers feeling frustrated and confused, and it took them significantly longer to buy what they were looking for. The scary part is that most customers don't make it to customer service to begin with. They log onto the website, get confused, and log right off.

Confusion is one of the biggest conversion killers out there.

If parents could go online and immediately go through a checklist or personalized quiz to help them narrow down the best bat for them, they'd be significantly more likely to buy. And if product names and descriptions were written for the customer, not for SEO, they'd feel more confident in their buying decision.

The Main Takeaway

Be careful what you emphasize. Customers will naturally search for the words, photos, and elements that confirm what they came here to find and fit into the mental category their brain has decided this purchase needs to follow. By knowing what they are—or are not—looking for and what language they use to express it, your company can meet them with the information they're expecting to find and shape their experience going forward.

AT A GLANCE

In the Discovery Phase, Your Customer Will Ask:

- Does this company understand my problem?
  - **Solution:** Carefully craft your first impression, which will create an anchor for the entire digital journey to follow.

- Does this company have a solution to my problem?
  - **Solution:** Conduct user testing to determine your customers' needs and the assumptions your customers come into the digital journey with so you can guide them to the right solution.

Principles at Play

- **Naïve realism:** The belief that our perception of the world reflects it precisely as it exists, untouched by personal biases or emotional influences.

- **Anchoring bias:** The idea that people overly rely on the first piece of information they encounter.

- **Availability heuristic:** Our tendency to use information that comes to mind quickly and easily when making decisions.

- **Confirmation bias:** The tendency to give more weight to information that aligns with preexisting beliefs.

Optimization Opportunities

- Homepage focus
- Website design
- Mobile layout
- Banner image
- Website imagery
- Website speed
- Copy
- Discounts and trial periods

In Summary

- When customers approach any type of digital experience, the initial information they encounter sets the tone for the entire shopping experience. Forming a strong first impression is vital.

- Customer decisions are often influenced by recent experiences or vivid memories.

- Understanding the assumptions that customers come into your experience with helps craft a positive digital experience off the bat and is an opportunity to redirect misconceptions.

- When customers have a preconceived notion about a product or company, they naturally zero in on information that confirms these beliefs, often ignoring contrary data.

- User testing is vital in understanding what your customers actually need and how they search for it.

Merely presenting information isn't sufficient for customers to find what they need. You must understand your website and products from the customer's perspective, recognizing gaps in information and navigation challenges that may not be obvious to someone familiar with the website.

3
INFORMATION-GATHERING

THIS IS THE POINT OF THE DIGITAL JOURNEY WHERE customers move beyond thinking, "Hmm, this might be what I'm looking for," to really digging into the details. During the Information-Gathering Phase, they move beyond casual window shopping and start a dedicated mission to find the information they need to make a purchasing decision.

This is the stage where customers begin to interact with your navigation bar, sift through different categories, explore software functionalities and features, delve into product descriptions, and scrutinize reviews. They're on a hunt for any piece of information that will confirm that your company has the solution they're looking for.

Where websites and apps often fall short is in assuming that simply putting the information out there means customers will find everything they need. Then, the sales and marketing teams get frustrated answering a thousand questions that, to them, are so clearly displayed on the website.

This is where you need to take a step back. Your team knows your products, services, and website backward and forward, which means they don't have the ability to see it the way your customers do. They can't see the gaps in information or the navigation issues that are too difficult to figure out unless the customer knows exactly where they're going.

Instead, you need to see the website through the customers' eyes.

When a customer starts on the Information-Gathering Phase, they go down a research path. Unless you want them bopping around the website aimlessly, moving from page to page and spiking your bounce rate through the roof, you need to create a clear, structured path to make sure they get the information they need, when they need it, with as little effort as possible.

To do that, you need to ask yourself:

INFORMATION-GATHERING

- What story are we trying to tell?
- What goal is the customer trying to accomplish?
- What do they need to know to accomplish that goal?

Once those directives are clear, it's time to strategize how to deliver that message. This strategy is less about the content and more about how the content is delivered.

Said another way, the facts don't matter. Or at least not nearly as much as companies give them credit for.

It doesn't matter what information is available on the website, it matters how it's presented—where it is, how it's said, at what point it appears on the screen, how easy it is to find, how it makes the customer feel, and how that impacts the feelings they built up about your product, service, or company along the way.

If you hide the information beneath layers of marketing speak, its value diminishes. If product descriptions are missing a key detail to differentiate them from their competitors, customers may just check out those competitors instead to get a feel for it themselves. It's your job to make sure your customers don't need to look anywhere else.

To form the best approach for your company and your customers, you need to experience your website like your customers do. They may ask:

- Can I find what I'm looking for?
- What do I need to know about this purchase?
- Who else has already bought this, and if I join them, what does that say about me?

QUESTION:
CAN I FIND WHAT I'M LOOKING FOR?

As customers delve into the digital marketplace, they are not just looking for a product—they are seeking a solution that aligns perfectly with their specific needs and integrates seamlessly into their "jobs to be done."[26] Customers need to be able to find that solution, even if they aren't exactly sure what the solution is yet.

Whether they're searching for a physical product or a service delivered through software, customers need to find the right fit—and they shouldn't have to wander aimlessly through the digital aisles to find it.

It is essential to create a clear and customizable browsing experience, recognizing that the search process is not just about finding a product but about discovering a solution that resonates with their unique requirements and goals.

---

26 https://hbr.org/2016/09/know-your-customers-jobs-to-be-done

INFORMATION-GATHERING

SOLUTION:
CREATE A CLEAR AND CUSTOMIZABLE BROWSING EXPERIENCE

Once customers have a general idea of what they're looking for, they need to be able to find it.

The problem is, companies often think of this search process in terms of helping customers find information that pertains to their company goals, not customer needs. Often, a website's navigational elements are packed with company-focused elements—like the blog or the founder's story—because someone on the marketing team advocated for it. While these elements are not without value, they shouldn't overshadow what the customer is actually looking for. Most customers are coming to solve a problem, not to learn more about the company's backstory.

Not only does this signal to the customer that this journey is not wholly focused on their needs—which is to find the product they need and convert—but leading with this information takes up the valuable space that readers are subconsciously more drawn to while hiding what they're actually looking for.

Let's dig into it.

Know How Your Customers Search

It's not enough to simply know what customers are searching for—you need to consider how they go about finding it.

At this stage the customer has a decent understanding that you have a good solution to their problem. But this question runs deeper—it's not "What *solution* am I looking for?" It's "What *quality* solution am I looking for?" The answer to that question isn't just going to affect what they buy—it's going to affect how they shop for it.

There are two types of shoppers: maximizers and satisficers.[27]

Maximizers need to know every single possible option available—not just on your website, but in the digital marketplace

---

27 Schwartz, B., Ward, A., Monterosso, J., Lyubomirsky, S., White, K., & Lehman, D. R. (2002). Maximizing versus satisficing: Happiness is a matter of choice. *Journal of Personality and Social Psychology*, 83(5), 1178-1197. https://doi.org/10.1037/0022-3514.83.5.1178

INFORMATION-GATHERING

as a whole. They need to make sure they make the very best decision possible. These are the shoppers who spend hours comparing nearly identical products or services side by side, who carefully weigh the value of every review and have so many tabs open in their browser that they can barely move between them.

Satisficers, on the other hand, want a solution to their problem—and they're only going to search for it only until they find a good option. Once they do, they're happy to hit "check out" or "sign up" and be done with it.

For satisficers, it's about getting the job done well. Good enough really is good enough. Maximizers need to find the very best.

Let's see the difference.

- **Maximizer:** Meet Alex, the CTO of a growing tech firm, who is the perfect example of a maximizer. On a mission to find the ideal project management tool, Alex leaves no stone unturned. He dedicates hours to exploring various options across the market, not just settling for the first few he comes across. He spends hours comparing products with similar features, diving deep into technical specifications, user reviews, and integration capabilities. Alex's decision-making process is exhaustive—he needs absolute assurance that the chosen software will not only meet but exceed his company's requirements for efficiency, scalability, and user-friendliness.

- **Satisficer:** Then there's Emma, a small business owner, who is a satisficer. Emma's on the lookout for straightforward, no-fuss accounting software to streamline her financial management. She's all about practicality—seeking a tool that's user-friendly and cost-effective and comes with solid customer support. Emma's search strategy is focused and efficient; she quickly evaluates a handful of top-rated software, homing in on their essential features. For her, the best software is one that meets her core needs and is easy to set up. Once she finds a software that ticks these boxes, she's ready to proceed with the purchase.

Some companies will naturally attract one type versus the other. Most will have a combination of the two. Companies that know their audience well typically know which way they lean, if they lean a particular way at all. You can create experiences that cater to both.

The secret is that the same experience that's going to instill trust in someone looking for "good enough" is also what's going to convince customers who need the best that your company really *is* the very best.

INFORMATION-GATHERING

To appeal to both at the same time, it's crucial to address *how* they seek out the information. That means moving beyond classifying users strictly as "maximizers" or "satisficers" to understanding the need to serve both "finders" and "searchers."

Take the example of a well-known bicycle company. A finder knows exactly what they're looking for—they've done their research and know the exact specifications of the bike they want. They're just deciding between two or three models. On the other hand, searchers are still in a learning phase. They know they need a bike and have a general idea of what they'll use it for, but they're not sure which bike meets those needs.

Ultimately, the goal remains the same: to help users find exactly what they need to find, in the way they naturally want to find it. Whether people are approaching the company as satisficers who are content with "good enough" or maximizers who want to scrutinize every detail, the aim is to streamline their experience and make it as seamless as possible. The more you analyze your customers' user data, the better you'll see the shopping patterns they create for themselves to fill that search need.

Let's say you notice that many users are opening multiple tabs; your initial thought might be that you need to redesign the website to prevent this. However, if you understand that your customers are maximizers who want to see all their options at once, you'll realize this behavior might actually be meeting their needs.

The question then becomes less about discouraging the opening of multiple tabs and more about finding easier ways for users to achieve their goals. Are they opening multiple tabs because the current website layout doesn't facilitate easy comparison of products? If that's the case, you might consider implementing features that make comparisons easier.

Understanding where your user profile leans could provide insights about what types of information to emphasize, although it shouldn't fundamentally change the overarching strategy. It's not as much about strictly catering to one type of customer or another—it's about creating tools that are robust (but simple) enough to give each customer the option to use them in exactly the way that works for them.

Account for Customers Who Know What They're Looking For as Well as Those Who Don't

The problem with traditional navigation is that it tends to skew toward people who already have a keyword in mind—the finders. Many companies think that if you come to their website, you should inherently know how to navigate through various drop-downs to get to the product or service you need. But what about those who don't yet know what they need? How do you make discoverability a cornerstone of the experience?

INFORMATION-GATHERING

In the realm of user experience design, it's important to consider "findability" versus "discoverability." Findability is purposeful. Discoverability happens by happy accident. Whether you have a large product inventory or a handful of curated service options, this distinction is critical. If customers know what they need, can they find it quickly? Or can they discover something new they didn't know they were looking for?

The answer to both of those questions should be yes.

Finders know exactly what feature or service they need. For instance, a finder might be a business manager looking for a tool that offers advanced data analytics with integration capabilities for their existing CRM system. They use search bars and direct navigation paths to quickly locate the specific software module or service page. They may filter through every feature they need, so they don't consider anything that doesn't meet their exact requirements. Your website must be equipped to lead these finders straight to their target with efficiency and precision.

Searchers, on the other hand, are exploratory in nature. They may recognize a need, like enhancing team collaboration, but they aren't sure which tool or feature suits them best. For these users, discoverability is key. When exploring different features and options, they may stumble upon solutions they hadn't originally considered. A searcher exploring might, for example, find themselves drawn to an interactive demo of a team collaboration tool they hadn't known existed but perfectly suits their needs.

In e-commerce, Etsy is a fantastic example of a platform built on findability and discoverability. The search bar pulls up exactly what you're looking for if you have a direction, and they show you every interpretation of that search imaginable. But they also leave room to discover the unexpected, whether it's along the path of searching for something else or as a starting point when you just want to look around. Their homepage lures you in with unique products, keeping you engaged as you stumble upon other items you didn't even know you wanted. It's like the e-commerce version of walking through a Target—you don't go there with a list; you let Target tell you what you need (unless you're much more commercially evolved than the rest of us).

When you implement elements of both, you work toward the goal of creating a culture of "engaged browsing" or "intentional browsing." This refers to a customer who is on a website with a specific purpose, whether it's to learn, find, or purchase.

Until you give them a reason, many customers will simply scroll without much engagement. Even if they originally came to the website with a purpose in mind, it's easy to lose sight of the original goal. We call that "zombie scrolling"—users who are looking just to look and not really seeing anything while they're at it.

INFORMATION-GATHERING

Why does this happen? Sometimes, customers intentionally come to a website just to look around. But most of the time, it means there was a falling-off point somewhere in the journey. They either lost focus or lost interest, and it's only a matter of time before they exit the page entirely.

It's your job to determine where that fall-off point is. Then, you can turn it into a moment to propel the customer forward.

But it's not necessarily a straightforward process. Understanding user behavior can significantly change how you read and interpret data. It's not simply about preventing a specific behavior—it's about understanding the reasons behind the behavior and adapting the user experience to better meet those needs.

> ### Let's See This in Action
>
> If you see customers scrolling straight through product descriptions, FAQ pages, or informational pages, there's a reason.
>
> Take a look at the page overall. If you see big chunks of text, it's likely that too many words are scaring off your customers.
>
> When category pages are text-heavy and image-dense, you create customers who zombie scroll through the screen—without actually taking in any of the info. Sure, you got them to the website, but those customers aren't going to buy.
>
> That's exactly what our team at The Good saw happen to one of our clients. So we looked at every product and asked, "What is the one thing customers really need to know about this product?" Not the ten things, not the five, not even the two—just one.

INFORMATION-GATHERING

But that doesn't mean the other information is irrelevant. Instead, we implemented a "read more" option. If customers needed more information to make a decision, they could find everything they needed easily and immediately. But if they didn't, they didn't have to sort through irrelevant details to find the information they really needed to know.

Remember, the Information-Gathering Phase isn't just a step on the digital journey—it's a journey in itself.

> If you bombard customers with all the information at once, they probably won't be able to take it all in. Instead, build on the information one piece at a time. What's the big picture? What information supports that big picture? Let them decide how much information they need (or don't need) instead of throwing it all out there at once.
>
> By replacing long product descriptions with one-line highlights on collections pages, customers were able to find (and buy) the right product quickly and easily—leading to $1.5 million in sales for this client.

Decide What Your Customers Need to See, Where, and in What Order

Just like we showed in the last example, the order in which you present information is just as important as the information itself. There are a few principles at play here that work together to influence this phenomenon:

- **Serial positioning effect:** The idea that we best remember the first and last items in a series and find it hard to remember the middle items.[28]

---

28   https://thedecisionlab.com/biases/serial-position-effect

INFORMATION-GATHERING

- **Primacy effect:** Our tendency to better remember information at the beginning of a series.[29]

- **Recency effect:** Our tendency to better remember information that was most recently told to us.[30]

The reason the serial positioning effect exists in the first place is because of primacy and recency. If we're likely to remember the first bit of information and we're likely to remember the most recent bit (the last item on a list), it stands to reason that the first and last positions are the most important when it comes to working within your customer's heuristic mental models. This also applies to items that are listed left to right. For example, when you list three items from left to right, people in the West will generally read the leftmost item first.

The key is to determine which items deserve those spots. Every company's criteria is going to be different, and it's going to change over time. It doesn't have to be your bestseller simply because it is the most likely to convert. You can leverage this principle to meet your company's goals, whatever they

---

29 Cherry, K. (n.d.). 4 Common Decision-Making Biases, Fallacies, and Errors. Retrieved from https://www.verywellmind.com/problems-in-decision-making-2795486
30 Cherry, K. (2020, April 9). What is the Recency Effect? Verywell Mind. https://www.verywellmind.com/the-recency-effect-4685058

might be, so long as they also meet the goals of the customer most likely to take action on that strategy. For example:

- **Goal:** Attract new customers.
- **Strategy:** The first slot holds the company's flagship service, an advanced cloud-based data analytics tool. This is their most sophisticated and sought-after service, likely to attract businesses looking for comprehensive data solutions. By positioning this service first, they ensure it captures the immediate attention of visitors.

  The second showcases the company's advanced services, which include significantly more features, customization, and support than the entry-level flagship version. These are targeted toward their returning customers who might be looking to expand their services.

  Third, the company features its latest offering, which expands on the capacities of the first two services but can be used separately as well. This may bring in customers who may not have needed the flagship products but see value in this unique offering.

Instead of telling your users where they need to be, think about guiding them through the path that best fits their needs. By applying these positioning tactics, you aren't tricking

INFORMATION-GATHERING

customers into buying something they don't want—you're putting what they need in the spot where they're most likely to look for it. If you can use these psychological phenomena to both reduce friction and achieve your business goals, then it's a win-win situation.

This tactic also helps returning customers remember where to look when searching for another purchase. If I asked you to buy twelve things from the grocery store, chances are you'd come back with the first two items, the last two items, and maybe one of the middle items. That's thanks to the serial position effect, which shows people have the greatest recall for the first and last items in a list.

You can take advantage of this effect in your navigation design. To drive visitors to your most important links, place them at the beginning and end of your menu. For instance, in e-commerce, if you know that your customers are bargain hunters, it would make sense to put a link to your "clearance" or "affordable solutions" category as the first navigation link. But if you have a high-end or highly specialized audience, it's going to hurt your company's reputation to lead off with sales.

Then, end strong. Many companies understand leading off with their most important link but then continue in order of decreasing importance from there, which typically puts something like the blog or founder's story at the bottom (which, for the record, don't belong in the navigation at all. They may help further the company, but in doing that, you're taking customers out of the digital journey completely with nothing but a hope that they'll make their way back).

So let's go back to the bargain hunters—in e-commerce, for example, it may be worth it to start with the sale and end with a "$50 and under category" link. Then, you can put pants, shirts, accessories, etc. in the middle, where users can find it if they want, but it's okay if they sit in the "messy middle," where people often lose focus or get overwhelmed.

The key here is to prioritize those anchor points—think about your primary product or service and how easily accessible it is for customers. Then, decide where those navigation points need to be. It's worth exploring other areas to figure out the optimal order. The goal is to identify places where reordering elements could significantly impact user behavior or sales, such as your navigation menu, category pages, filters, product descriptions, and calls to action.

These principles even impact your product and service descriptions. Eye-tracking heat maps often reveal that people start by reading the first sentence or a few words and then quickly lose interest. They either skim just the left side, reading the first couple of words of that large paragraph, or they read the first sentence and then jump to the last one. They're basically looking at the beginning and the end and skipping over the middle.

INFORMATION-GATHERING

The reason for this behavior is that people seek context. They want to quickly determine whether the content applies to them. If it does, they're more likely to read the entire text. But more often than not, they're not going to read every word—they're skimming to find what's important to them.

When writing product descriptions, here are a few good rules of thumb:

- **Immediately delete the first sentence of your first draft.** Maybe even the second. Don't waste time introducing your product again—get straight to the information your users need to know.

- **Keep it simple.** Lead with the attributes that your users absolutely need to know in order to make a decision. Don't bury the details that are most likely to convert in a slew of other information that may or may not be relevant to each particular user.

- **Utilize "read more" drop-down menus.** There are going to be users who do want to get into the nitty-gritty of the product information. This drop-down menu gives these users the option to find what they need without having to leave the page, without bombarding other users with the information they don't care to see.

Let's See This in Action

The reason companies need to carefully choose the order in which they present options isn't just to get attention—it's to reduce friction in the shopping experience.

Every small roadblock in a customer's path makes them more and more likely to give up entirely. Finding those roadblocks gives your company a chance to not just keep customers on the website but propel them further down the purchasing funnel.

For one client of The Good's, we found that roadblock within the navigation. This company came to us with a menu listing items in their header navigation like Shop, Connect, Discover, Rewards, etc. On top of that, the

INFORMATION-GATHERING

Shop drop-down listed product categorization with labels like Collections and Themes.

Logical, right? But it wasn't working. This categorization wasn't clear enough for new visitors.

So we dove into what customers were actually looking for and found a handful of specific categories that bar and above received the most clicks.

This told us which links customers put in the effort to find—which is impressive because the vast majority of customers will abandon a website at the first roadblock rather than spend time figuring it out (no matter how minorly inconvenient it is). If enough people put in the effort to unscroll a navigation bar to find what they're looking for, we could only imagine how many logged off the website entirely when what they really wanted was only a click away—they just didn't know it.

So we tested adding top product categories into the top-level menu navigation for better browsing opportunities, product catalog awareness, and increased transactions.

Our variant increased conversion rates over the control, resulting in over $1.5 million of revenue gains.

Personalize Your Filters

Filters help visitors quickly sort through your product collection to zone in on the items they need.

Think of filters like tightly focused, super-helpful search engines that guide visitors deeper into your products, services, or features. If a customer wants a red T-shirt, for instance, a proper filter would let them view a collection of products that meet the attributes of "T-shirt" and "red." If a customer is looking for cloud storage solutions with specific attributes like "encryption" and "large file support," the right filters could quickly display all relevant options. This approach streamlines the process and allows customers to explore options more intentionally.

The beauty here is that when done right, filters can be applied in any number of ways to help the user achieve their goals. Maximizers are able to see all options within their search parameters, clearing out the clutter so they can look at everything at once and making it much easier to make the best decision. For satisficers, all they have to do is pick their criteria and find an option that works. Finders can execute the search they came here for. Discovers can explore their general options before looking into specific products, services, or features.

No matter how a customer shops, when done right, filters streamline the process in a way that works with the customer's search style.

But the "when done right" is where it gets tricky.

The more specific the options, the more likely the customer will be able to quickly find what they need and convert. But this puts a higher burden on you to make the filtering experience as easy as possible. The more options available, the more important it is to organize them in a way that is easy to navigate. If your user's most important attribute is hidden in the middle, it's going to be harder to find.

If you need to optimize your company's filters, here are a few things to keep in mind:

- **Select your filters carefully.** While there are some best practices, each option should be specific to your company and your customers. Here is a good place to start:
    - Company (but not if everything you sell is from the same company)
    - Price
    - User ratings
    - Features
    - Color
    - Material
    - Size
    - Popularity
    - Promotion (e.g., new, featured, or on sale)

- **Keep your product filters simple.** The easier they are to identify and operate, the better your visitors will like them. Use filters that are reasonable and intuitive. Don't create extra filters just to have more. It's fine if you only need three or four.

- **Test as you develop.** A filtering system that makes sense to the design team may not play well in front of a live audience. Test early, and test often. The sooner you catch problems, the easier they are to solve.

- **Consider your products from the viewpoint of your customers.** What information do your customers need in order to make a wise purchasing decision? What do your customers desire? Do they favor certain colors over others? Do you see seasonal jumps in certain product attributes? Leverage your data as much as possible to discover what your customers care about.

- **Use the words your customers use to describe your products or services.** Avoid industry jargon. Your filters should be absolutely practical and common sense. Don't make customers struggle to find exactly what they're looking for.

- **Look for higher-level groups of attributes for busy categories.** Items and services might be grouped around themes or use cases. These are called thematic filters, and they help visitors quickly home in on the desired product, especially when they're in the early stages of their buying journey and don't know exactly what they need.

- **Give visual confirmation of active filters.** Make sure the visitor receives visual confirmation that your filter navigation is active. For optimum user experience, the current filter state (waiting for input, working, and output delivered) should be readily apparent.

- **Consider using different filters for mobile and desktop.** In some cases, it makes sense to offer different filters on mobile and desktop devices. For instance, you may want a stripped-down version for your mobile website that doesn't include rarely used options.[31]

---

31   https://thegood.com/insights/ecommerce-product-filters/

Let's See This in Action

For one client, they had a good handle on the top of filters they needed and which order they should appear. But customers still had a hard time using the filters. Why? Because once they chose a filter—especially on mobile—it disappeared. That was fine if they were just using one or two filters, but if they had a specific set of filters on, it could be easy to forget what they were looking for in the first place. And if they wanted to change the filters? Time to start over—it was going to disrupt the process entirely.

So we added the customer's selected filters to the top of the product menu, which brought in a $905,000 revenue increase for our client—all because it was easier

INFORMATION-GATHERING

for customers to search for the products they really needed.

Why does this work?

Having a section at the top of your product page that clearly shows the filters your customers have selected means they always know exactly what they're looking at—and reminds them of what they need to buy.

When filters stay hidden in the sidebar, it forces customers to go back into the navigation, scroll through all available filters, and manually check or uncheck what they want to look at. This significantly slows down shopping, especially for people visiting your website on mobile.

The longer it takes for customers to find what they want, the less likely they are to buy.

Your website's purpose is to help customers find the products and services they need when they need them. This small change does just that.

Hone Your Category Pages

When it comes to category pages, the question isn't, "How do these items go together?" It's "How do these items or services meet the customers' needs?"

Every page on your website or app has a purpose. Your job is to know what that purpose is so the visitor will glide effortlessly through to the next step on the path leading from discovery to purchase.

Understanding the "jobs to be done" framework is crucial in this context.

This framework shifts the focus from "What does the product or service do" to "What does the customer need the product or service to do?" Instead of focusing on its actual attributes, it focuses on what the customer is trying to accomplish.

McDonald's milkshakes are a perfect example.

McDonald's initially refused to offer milkshakes during breakfast hours. They couldn't understand why people would want a milkshake in the morning. It's a dessert, so surely it must be better for dinner—or even lunch if you really want to treat yourself.

Then, McDonald's brought in a researcher to help refine their breakfast menu due to the unusual requests they were getting. After extensive consumer interviews, they came to terms with the fact that people wanted milkshakes in the morning. Instead of continuing to say no, they asked why. What they found was that many people ordered milkshakes because it was a quick and easy option to get protein and stay full until lunch, all while sipping on the go. To them, it wasn't just a dessert. This "job to be done" for a morning milkshake was very different from what the McDonald's team initially assumed, and this insight significantly informed their breakfast menu going forward.[32]

This is exactly the approach companies need to take with their category pages.

Because the first option is the most likely to be clicked, this is where you put your category that answers the biggest need your customers come to you for. If a major shoe brand has a category for winter shoe staples and another for shoe maintenance and repair, they're going to lead off with the shoe staples. Leather cleaner is a valuable product, but not without shoes to clean. If a SaaS company offers both comprehensive CRM systems and supplementary data analysis tools, the CRM systems should take precedence. The supplementary tools aren't useful if the base service is not in place.

For category pages, it can be as simple as just asking the visitor what they are shopping for. In e-commerce, if they're shopping for tables, do they want to see all their options—from bedside tables to entryway consoles—or are they specifically looking for kitchen tables? Are your users typically established older adults looking to invest in one long-term purchase, or are they new homeowners who need to fill the entire kitchen? This is where a "Shop Kitchen" category could

---

32   https://hbswk.hbs.edu/item/clay-christensens-milkshake-marketing

be a great option so they can find what they're specifically looking for while also filling additional needs without additional effort.

If you're a SaaS that offers a wide variety of services you can categorize by industry, putting all of the project management systems in one category while the invoicing platforms are in another. Or you can categorize by company type. A small business owner, for example, may need both a project management system and an invoicing platform—plus a customer relationship management software while you're at it.

If you don't already know, it's time for some user testing to find out. For most products or services, a handful of targeted questions can guide someone toward a recommendation quite efficiently, especially when you have a narrow range of options to begin with.

Once you understand what your customers are looking to accomplish, you can tailor your categories around those specific needs. This makes the decision-making process easier, helping them navigate through choices without making them feel overwhelmed.

A great category page should include these essential elements:[33]

---

33   https://thegood.com/insights/product-category-page/

INFORMATION-GATHERING

- **Product or service category name:** Every case is unique, but you'll typically want to use the product or service name your prospects use for the category or individual product, not a branded or esoteric name. That not only helps the customer understand your categories—it also helps the search engines key in on them.

- **Category image:** Make sure this is a generic image representing the entire category. Make it a hero image of the product by itself or a clear screenshot of the service platform. The simpler and more obvious, the better. The images should be consistent in size but should stand out as different from one another. Guide the eyes, and the minds will follow.

- **Category price ranges:** It's usually better to give shoppers the option of selecting the price range that best fits their budget than to lead them to a solution way out of range. The sales department may argue the strategy, but we've found well-informed customers are usually the happiest customers.

- **Category ratings:** Ratings and reviews can be invaluable. Here's where the sales department may indeed have valuable input. Shoppers will often bump up in price for higher quality, and quality is typically reflected in the reviews.

- **Featured items or services:** Don't confuse the shopper, but don't be afraid to make suggestions either. Depending on placement, featured items or services can draw the eye and lead to larger sales.

Let's See This in Action

When one of our clients at The Good came to us for a conversion rate optimization strategy that would work particularly well over the holidays, we immediately jumped to the "jobs to be done" framework.

This well-known outdoor and industry footwear brand with a very in-the-know audience proved a unique challenge in terms of gifting. The products themselves did make excellent gifts, but because their options were so specialized, it was difficult for customers to figure out exactly what their loved ones might want.

Heat maps showed that while customers interacted with product tiles, they were less likely to click them to learn more about the products.

So instead of guiding them toward one option

or another, we chose to simply tell them. By adding a "top gift" product flag, we helped customers feel encouraged to learn more about this curated selection and ultimately make a purchase—resulting in a 5.61 percent conversion rate increase.

Streamline Your Navigation Menu

You can argue that if there's one place where psychological principles are going to impact your website or app's digital journey, it's going to be in the navigation. This is how customers get from place to place, so you want to make sure they're naturally heading in the right direction.

There are three main navigation methods:

- **Horizontal navigation bar**: The header navigation bar is the most common type of website navigation. These top-level links are displayed horizontally, side by side, in the header. They usually include the most important pages of the website—the pages you know users will need access to early.

- **Main navigation menu:** This is where users go when they need to find a product or service—it is the main navigation used in the conversion process. This is where setups vary. The most popular, and useful, options include:
    - **Drop-down navigation menu:** Drop-down menus are tiers of navigation that keep your content organized. The most important links are used as top-level navigation. When a user clicks or hovers over a link, a menu expands with more links. There could be several tiers of drop-down menus.
    - **Hamburger navigation menu:** You see a lot of this menu on mobile websites and apps. Instead of listing any of the top-level links on the page, everything is hidden behind the button. Tapping the icon reveals the menu. Tapping a menu item expands the menu further, revealing additional options.
    - **Vertical sidebar navigation menu:** A vertical sidebar menu is a list of links stacked on top of each other, positioned on either side of the page.

This format is typically better for websites that have too much top-level navigation for a horizontal menu. The downside is that they eat up a lot of page real estate. Sidebar navigation is also useful on interior pages where users need more specific options and internal linking to facilitate their shopping experience.

- **Footer navigation:** The footer navigation is a "catch-all" spot. It typically contains links to most pages on the website (though individual blog posts and product pages aren't usually included). While you shouldn't cram a link to everything into your footer menu, it's smart to offer lots of options.[34]

As you see, the main navigation is the most important area for the conversion process and the one that stands to benefit the most from implementing the principles of primacy and recency effects.

If you want customers to find the product or service they're actually looking for as quickly as possible, a great place to start is simply by reducing the number of options they have to see. The rule of thumb is to reduce your top-level navigation links

---

34 https://thegood.com/insights/website-navigation/

to seven. More information to process means a bigger cognitive load on your users. After seven items, people stop reading and start scanning, which means they potentially miss information that could be important to them.

Let's See This in Action

Consider a paint company we worked with a few years back whose navigation menu was less than optimal.

The first drop-down menu on their website had a tab labeled "Shop," and under that, they listed various products. Astonishingly, paint, which constituted over 90 percent of their sales, was buried deep within this list. They also offered other products like primer, floor varnish, and swatches, but paint was undoubtedly their bread and butter.

While some might argue that this setup gave visibility to their other products, in reality, it added unnecessary friction to the customer journey for their primary product—paint. On top of that, the core and supplemental offerings were buried in company-focused elements like "Learn," "Blog," "Project Certifications," and "FAQ,"—all things that may be useful information in another context but don't help the customer make a purchase.

INFORMATION-GATHERING

Following the primacy and recency principles, we knew we needed to anchor the core product offerings at the beginning and the end. So we put "Paint" as the first item listed and "Samples" as the last. This approach serves two types of customers: those with high intent, who are immediately presented with the main product, and those in an exploratory phase. Placing "Samples" last helps guide these exploratory customers down the path of purchase readiness. Offering samples can be particularly effective since many companies find that their conversion rate from a sample to an actual purchase is very high. It becomes the next best action to make a purchase.

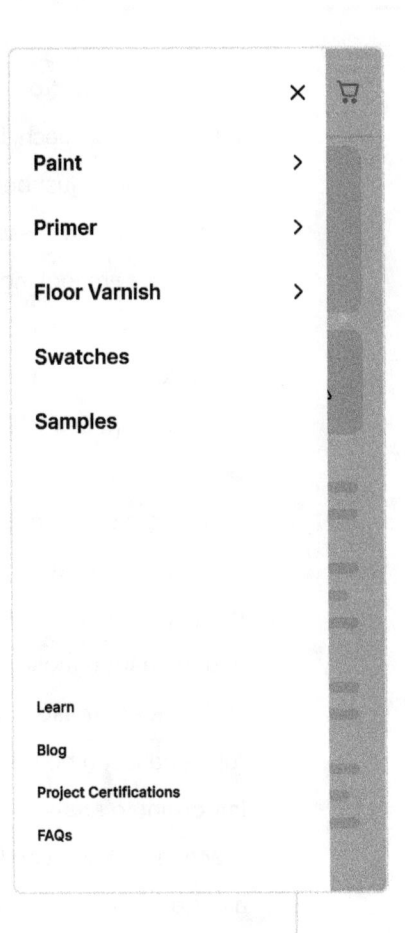

This example also underscores the difference between what you might assume to be the correct order and what actually works best. You might think that listing items from most to least important would be the logical approach, but that's not necessarily the case.

Remember: just because it's logical doesn't mean it's true. You can't assume you know what your customers are looking for—you have to review the data.

Now, just because you know what their customers want doesn't mean that you absolutely have to change it up. Especially if your company is looking to push a specific product, it may be useful to use those prime-time spots for something other than the more tried-and-true navigation items. If you sell a lot of paint but you want to sell varnish, it might be worthwhile for you to change those around.

There are reasons you might want to break that rule, and it can be done well. This is where it's important to know the principles behind these "best practices." When you know why customers behave a certain way, it gives you the information you need to know when to work with it and when to break the mold—and do it well.

Optimize Your Search Bar

Visitors who use the search box typically have the intent to buy or act. So if your customer is asking, "Can I find what I'm

INFORMATION-GATHERING

looking for?" you'd better believe they're going to head to the search bar at some point.[35]

Search bars are one of the most powerful revenue-generating tools in your arsenal. Website managers typically understand the importance of site search to sales, but many assume that simply incorporating the function into your website or app is enough. If you want to ensure that each query consistently and reliably pulls the results your customer needs, you need to optimize.

The better your search engine is at returning the desired results, the more likely your visitors are to make a purchase. In fact, visitors who take advantage of the site search tool convert at almost double the rate of those who don't.[36]

Optimized search functions show results leading to the best solution to that query, which is a great way to take advantage of the primacy effect. Just as no one searches past the first page on Google, your customers assume if you have a good solution to their search, it's going to show up first. So make sure it does.

There are a few best practices that will help customers get where they need to be:

---

35 https://thegood.com/insights/website-navigation
36 https://econsultancy.com/is-site-search-less-important-for-niche-retailers

- **Make site search highly visible.** If your visitors don't notice or can't find your website search function, they won't be able to use it. There are plenty of other ways to conserve space without sacrificing site search visibility. Site search should be above the fold, and it should stand out like a fire extinguisher at the race track.

- **Go strong on autocomplete and error correction.** Don't expect shoppers to know product or service names, how to spell them, or even what they're looking for in the first place. Your site search function should be smart enough to anticipate where the search is headed and begin listing suggestions while the shopper is still typing.

- **Never allow site search to hit a dead end.** "Product/service not found" should never be in your website's vocabulary. No matter what a customer is looking for, they should always be able to find something. You know who never leaves searchers completely stranded? Amazon, where even the wildest of searches yield 200 results—and that's exactly where your customers are going to go.

- **Optimize for the terms your customers use, not for industry jargon.** You want your search engine to serve your customers, not your competitors. Speak the language your customers speak. If outdoor gear is your specialty, those "hook and loop" fasteners are commonly known as "Velcro closures." If your customers use the word "vest," you should be using it too.

- **Consult your analytics data for website search insight.** Monitor analytics data related to on-site search. You'll get a clear picture of the terms your visitors are entering in the search box, the results they're being shown, how identified segments of your audience use search differently, and more.

- **Provide search access to more than your products only.** Your visitors may still be researching a category of products instead of seeking to buy a particular product. Make sure your on-site search function provides the desired results. It should index size charts, category overviews, technical data, and other relevant resources—not just product names. Include SKUs in search.

- **Provide searchers with abundant filtering options.** Let them search the way they want to search. Some will want to filter by department or topic. Others will be interested in the best reviews. Configure your site search engine with an eye toward usability and customer experience. Happy shoppers buy more per visit and come back sooner to shop again.

- **Don't overlook the importance of metadata.** The tags, titles, and descriptions you apply to your products are the bread and butter of search. Make sure the terms (keywords) your customers most often use to describe each product are included in that metadata.

- **Take special care to configure mobile search functions.** If it's difficult for mobile users to click on that little magnifying glass or otherwise interact with your mobile search box, you're going to lose business. Maybe buttons work better than links. Maybe you want the search box to open automatically. Whatever you decide to do in the way of on-site search design, prove the effectiveness of your choices with extensive user testing.[37]

The easier it is for your customers to find the information or products they need, the quicker they can move into the next phase of the digital journey.

---

37   https://thegood.com/insights/ecommerce-site-search/

Let's See This in Action

When working with a client in the car industry, we found a roadblock in their customer's search process. Heat maps showed that the users were drawn to the search function but were unsure where to go when their options looked too similar.

To us, that showed that somewhere along the search process, the website stopped guiding the digital experience. Our hypothesis: changing up the search function to include more visual cues would encourage visitors to use the vehicle search and increase conversions.

So we tested two ideas:

1. Show contrast so it draws users visually to a light placeholder text that gets bold when you select an item.
2. Use some subtle visual design language to

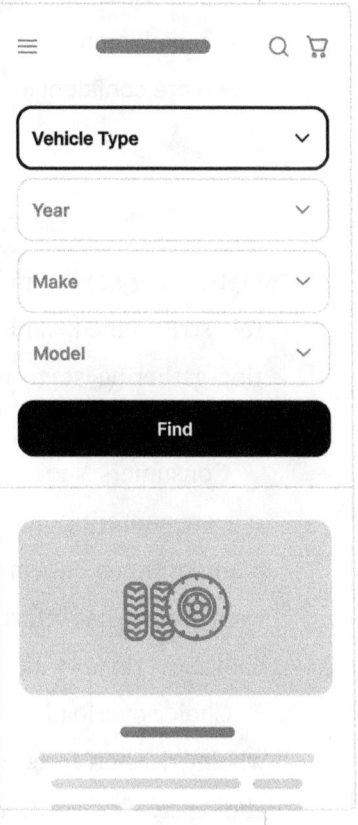

> really make the search pop for users so if they use it, they will confidently find a product.
>
> After testing them both, we had a clear winner. Using the visual cues in option 1, users were able to search more easily and intuitively, making them feel more confident in their purchase decisions.

Avoid Overwhelm

Optimizing search and navigation creates an extremely helpful, streamlined, and personalized path from the information-gathering stage to the decision-making stage. But it's possible to go too far.

Consumers should be able to find what they need without encountering undue difficulty. But when faced with too many options, customers can feel overwhelmed before they even know what information they're looking at. That's where choice overload comes into play.

Choice overload is the tendency for people to become overwhelmed when faced with too many options.[38] This is a psychological element that will impact customers at every stage

---

38   Chernev, A., Böckenholt, U., & Goodman, J. (2015). Choice overload: A conceptual review and meta-analysis. Journal of Consumer Psychology, 25(2), 333-358. https://doi.org/10.1016/j.jcps.2014.08.002

INFORMATION-GATHERING

of the process (so you can bet we'll be coming back to this when we talk about the Decision-Making Phase.)

In the Information-Gathering Phase, though, choice overload can keep customers from finding the information they need to make a decision simply because there is too much to sort through. If navigation is confusing, if there are too many filters to sort through, or if it looks like there are a million categories, customers may feel overwhelmed from the start.

And then they're going to log off.

Essentially, choice overload results from the company getting in the way of helping the consumer. Your goal should be to assist people in quickly identifying the products or services that can address their needs or pains. Failing to focus on the consumer leads to navigation centered around what the company cares about, insufficient filtering options, or irrelevant product showcases that aren't going to prompt anyone to buy.

While it might seem like consumers are always on a mission to get in and get out, that's not always the case. Especially in this phase, many customers want to spend time considering their options and the qualities associated with each. For these consumers, what's needed is a "guided path," the term we often use as the opposite of choice overload. People sometimes require a little "life concierge" to navigate their choices, much like when you go into a high-end store like Saks or Nordstrom. The sales associate not only assists you with what you pick but also suggests additional items that suit your taste and budget.

This makes it crucial to carefully consider every element of your website that may contribute to choice overwhelm while searching for information:

- **Filters**: Especially for companies with large portfolios, the lack of a clear and obvious way to filter results is going to create frustration right off the bat. You get choice overload when you have a bunch of results that you can't intuitively whittle away to the ones that are most relevant to your use case.

- **Category pages**: Categories can be an excellent way to guide users on a jobs-to-be-done framework. But when there are too many options, it makes their solution harder to find (or even makes them question what solution they were seeking in the first place.)

- **Navigation menu**: You have to ask, "Can I go through our navigation and understand where I need to go next to solve my pain point or need, or do I have to wander through all the aisles of our website and just look up and down until I find the product I want?" Anything that doesn't specifically

INFORMATION-GATHERING

apply to the purchasing process (like company history) doesn't belong here. Try to whittle down the options—anything more than seven is going to cause overwhelm.

Let's See This in Action

Let's be clear—an overwhelmed customer is not likely to convert.

A client came to us with a problem: because they had so many offerings, customers felt content fatigue before they found what they needed.

Using heat maps and sessions recordings, we found two things:

1. Users showed high directness on the side filters.
    - **What did that mean for the user?** Users were able to find what they needed using the side filters.
    - **What did that show us?** Customers already expected to search using these filters—they just needed a few tweaks.

2. On category pages, users were scrolling up and down.
    - **What did that mean for the user?** It was difficult to find what they needed, and with so much content, they didn't know what direction to follow.
    - **What did that show us?** The need for better filtering, content hierarchy, etc.

INFORMATION-GATHERING

> By simplifying the sidebar and allowing for more search customization in the category pages, we were able to better direct customers to the products they were looking for—which meant a conversion rate increase of 10.8 percent for the client.

Putting It All Together

Your company's browsing experience is at the core of the digital journey. The easier, more streamlined, and more guided it is, the better the digital experience will be.

There are a number of ways to optimize this experience:

- Account for customers who know what they're looking for as well as those who don't
- Decide what your customers need to see, where, and in what order
- Personalize your filters
- Hone your category pages
- Streamline your navigation menu
- Optimize your search bar
- Avoid overwhelm

QUESTION:
WHAT DO I NEED TO KNOW ABOUT THIS PURCHASE?

Consumers are more discerning than ever, given the multitude of options available. And in the age of online shopping, when they can't touch or feel the products, getting clear, concise, and adequate information is even more critical. Or, in the case of SaaS, the product isn't tangible at all. They want to ensure they're making a choice that offers not just utility but also quality and longevity because making a wrong choice can be both frustrating and costly.

The key to this question is, "What do I need to know about this purchase?" Not "What would I like to know?" Not "What's interesting to know?" And definitely not, "What does the company want me to know?"

Filling that need often means offering less information, not more. Instead of throwing every piece of information out there all at once for the customer to sort through, distill it down to the key details. Give them what they need, when they need it, and leave it there.

What they need:

- **Product or service specifications:** What are the key features or dimensions?

- **Customer reviews:** What experiences have other people had?
- **Price comparison:** Is it competitively priced?
- **Return or refund policy:** What if it doesn't meet expectations?
- **Compatibility:** Will it work with other products or software they own?
- **Company reputation:** Is the manufacturer trustworthy?
- **Shipping information (in e-commerce):** How quickly will it arrive?

It's essential that the information presented during this phase is clear and concise and that it directly addresses the questions shoppers actually have. Filling the screen with what marketing teams think is useful—but is actually not—can muddle the process, slow down the decision-making, and even jeopardize the sale.

Customers don't need all the details—they need the right details. When there's one thing that sets the product or service apart from the others but you include a list of ten key features, you turn what was the defining attribute into nothing more than a bullet point.

You don't need to delete valuable information or details that could leave some of your customers with questions they can't find the answers to. Instead, you distill the initial information customers see to the most essential details and then give them an option to seek more information if they need it.

Once you narrow down what your customers need to know off the bat, you can move on to making sure they get the message loud and clear.

The goal is to make them want more information so that they choose to explore additional details on their own, creating a much more engaged and intentional path to purchase. Or, better yet, to make the message so clear that they don't need anything more, and they're ready to make a decision.

To do that, you'll need to:

- Leave nothing up to interpretation.
- Say it again…and again…and again.

SOLUTION 1:
LEAVE NOTHING UP TO INTERPRETATION

The idea that "facts are facts" may be true in terms of accuracy, but along the digital journey, it's not enough. The way you choose to share the information has a significant impact on how your customer interprets it. This is known as the framing effect, which is the idea that our decisions are influenced by the way information is presented. [39]

Everything your company puts out there either adds value or takes it away, based on how you present information—even if the information itself is objectively valuable.

Let's take percentages, for example. You can either say your product or service has a "10 percent failure rate" or "90 percent success rate." It's a no-brainer—people are going to want the option that offers a 90 percent success rate.

At the end of the day, it means the exact same thing. But if you put them side by side, you know exactly which one you'd pick. Mistakes in how you frame the information can deter potential buyers even if the product is generally well suited for their needs.

From product descriptions to customer reviews, every element should be framed in a way that guides the consumer positively through their shopping journey (while staying truthful).

Framing can impact the way you set up any number of areas on your website.

Product or Service Descriptions

So many companies assume that if they want to tell their customers about the product or service's various features, they need

---

[39] Tversky, A., & Kahneman, D. (1985). The Framing of Decisions and the Psychology of Choice. *Behavioral Decision Making*, 25-41. doi:10.1007/978-1-4613-2391-4_2

to tell them *everything*. So they load up the page with mountains of text, highlighting every single detail because, of course, that should be helpful when gathering information, right? Instead, the customer just feels like they're shouting at them.

Unsurprisingly, most customers aren't interested in being yelled at, and they're even less interested in being told what should be important to them—especially if it really isn't. Instead, they want to feel like you already know what's important to them and you're addressing it without them having to ask.

When you apply framing to your product descriptions, you create a path that helps the customer understand exactly why they should choose that product, without them having to decipher all the information and piece it together on their own.

Understanding your target audience is essential for effective framing. Let's take cookware, for example. If your primary customers are chefs, they'll likely require detailed information about the product. However, if you're targeting young home cooks who live in small city apartments with limited kitchen space, their needs will differ significantly. They're looking for a single, versatile knife that can get the job done quickly. They don't think that they need a knife that can scale fish, peel fruit, and carve meat. What they care about is having just one knife so their kitchen isn't cluttered.

Ultimately, it's the same multifunction knife. One description says what the knife does and leaves it to the customer to

INFORMATION-GATHERING

determine whether that's what they need. One gets the heart of their pain point, leaving nothing up to interpretation that this is the solution they were looking for.

It all boils down to knowing who your audience is and tailoring your message accordingly. Are you targeting chefs who may need a range of specialized knives, or are you aiming to simplify the life of a parent who just wants to get through cooking dinner as efficiently as possible? Once you have that clarity, you can frame your product or service in a way that resonates with your specific customer base.

Said another way, make one clear argument. Information is important, but every piece of information should lead to one clear message. The more points you try to make, the less powerful each of those points becomes.

Then, present your argument accordingly.

Let's See This in Action

Sometimes, you can sell more by promoting less.

Years ago, our team worked with a company that sold at-home meal kits. Under each recipe, they listed the individual ingredients as a group, and while this layout initially brought an increase in their conversion rate, the data showed that it still wasn't exactly right.

So we decided to test another layout to see which one customers responded to best. Instead of grouping all the ingredients together like a digital shopping cart, we changed the layout to place more focus on each single ingredient.

This increased their conversion rate by 31 percent.

Why? Because this client knew their audience, and they framed the information accordingly. Think about this: if you're someone who goes to the grocery store once a week to stock up, how often do you really calculate the total cost of each individual meal? Likely, not often.

So when customers could suddenly see the cost per meal, it often felt more expensive than they were used to. It may not have been, but it felt like that. After all, if you choose to eat dinner at home instead of going to a restaurant, that meal feels free, right?

Creating a larger focus on each individual ingredient meant customers felt like they were gaining even more value. One recipe kit may not feel like it's worth the price. But a cartful of ingredients (that just so happen to work perfectly together to make one recipe)? That feels like a good purchase.

It was all about the perception. Framing the recipe as something bigger than just one meal gave customers a higher confidence in their purchase.

Social Proof

Making a purchase is a journey, not a one-off event. The digital experience should make buying products or services a logical step on that journey. If you lead your prospects to a "no-brainer" decision that feels right, they'll become happy buyers who buy from you again and share that good feeling with their friends.[40]

Social proof is a great way to frame your product as the no-brainer choice. It was a no-brainer for others, so it must be for you.

---

[40] https://thegood.com/insights/social-proof

Getting others to promote you is infinitely more effective than just promoting yourself, which makes social proof one of the most powerful tools you can apply to your website.

Social proof is the idea that people are likely to emulate the behaviors of others in order to reflect the correct behaviors in similar situations. This effect is heightened when we lack sufficient information to make informed decisions on our own. We presume that other people have more knowledge about the situation, so we should follow their lead. That's why it's also referred to as herd mentality.[41]

People utilize social proof every day while making decisions, often without even noticing. If a restaurant is busy, you assume it's good. If it's empty, you think there must be a reason.

The same goes for the digital journey. If a friend suggests a product or service, you're more likely to buy it than if you just saw it in the store on your own. Why? Because someone has vouched for it, so it must be good.

In the context of framing, data-driven social proof can be a huge asset or a hidden liability. We see it all the time—if we asked if Colgate is a good toothpaste, you may immediately know that "four out of five dentists recommend Colgate."

---

[41] https://www.amazon.com/Influence-Psychology-Persuasion-Robert-Cialdini/dp/006124189X

INFORMATION-GATHERING

Whether it's true or not is up for debate, but it's a powerful example for how to frame the numbers in your favor.

When you say that "four out of five dentists recommend Colgate," that sounds pretty good, right? Most of the dentists are in agreement—who cares what that one outlier thinks? But here's the thing—that interpretation changes the bigger the numbers get. You could say that "eighty out of one hundred dentists recommend Colgate." Mathematically, it's the same ratio. But all of a sudden you're now thinking, "Who are these twenty dentists who don't recommend Colgate?"

The data doesn't change—your interpretation of it does.

When it comes to framing social proof, smaller numbers feel better than bigger ones. It's easier for customers to grasp, which makes it feel more credible. Realistically, eighty out of one hundred is a larger sample size, which should show a higher degree of confidence—after all, there are thousands of dentists in the world, so it wouldn't be hard to find four who recommend Colgate. But it feels better, so it must be better.[42]

On the other hand, the reverse works just as well. When using data that puts you in the minority, bigger numbers make the gap less glaring than smaller ones. You could say that one out of four consumers prefer paper straws to plastic—which

---

[42] https://www.ted.com/talks/niro_sivanathan_the_counterintuitive_way_to_be_more_persuasive

doesn't feel like a lot. Or you could say that 100 out of 400 consumers prefer paper straws to plastic. Sure, it's not the majority, but 100 people prefer it, and that's no small thing.

It also matters how you present the numbers. You could say four out of five, or you could say 80 percent. Social proof is meant to pull people in, to make them a part of a group (ideally, the group with the right opinion.) It's much easier for customers to see themselves as a part of a group of five. But as a percentage? No one wants to be just a statistic.

When framing data-driven social proof, remember:

- When you're in the majority, use smaller numbers.
- When you're in the minority, use bigger numbers.
- Fractions are more influential than percentages.

This doesn't mean you should manipulate the numbers—whatever claim you make, make sure it's true. But if it's a claim you're willing to put on your website, it's probably likely that the numbers add value to your product. So you want to make sure to frame those numbers in a way that gives your customers a feeling, not just a fact.

It's not about allowing people to form assumptions based on irrelevant or misleading data. The aim is to provide reliable information that genuinely helps the customer, rather than perpetuating misconceptions that aren't supported by data.

INFORMATION-GATHERING

Let's See This in Action

Number-driven social proof came in handy for one of our clients at The Good who is on a mission to improve the quality of our drinking water.

A recent United States Geological Survey revealed that 45 percent of Americans may have contaminated water—specifically highlighting PFAS (a class of 12,000 different contaminants that can enter the water supply).[43]

Unsurprisingly, customer interactions proved that this was a point of major concern for this brand's target audience. So we leaned into the numbers.

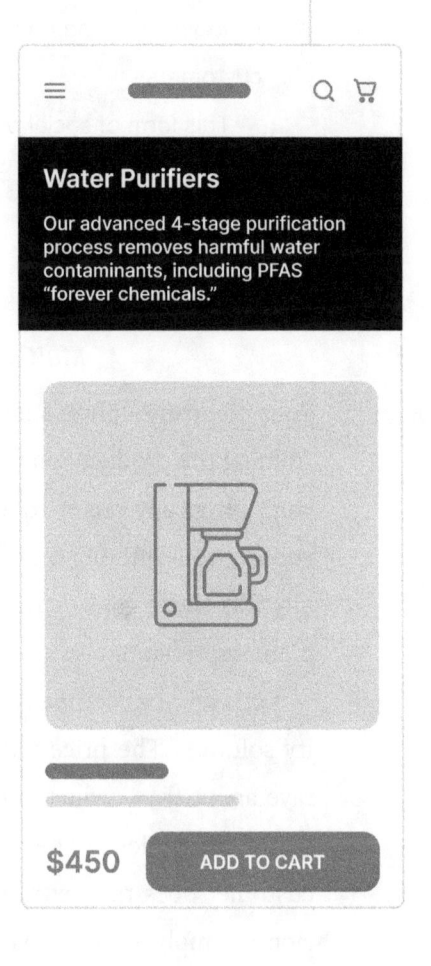

Water Purifiers

Our advanced 4-stage purification process removes harmful water contaminants, including PFAS "forever chemicals."

$450 ADD TO CART

---

[43] https://www.usgs.gov/news/national-news-release/tap-water-study-detects-pfas-forever-chemicals-across-us

> Using data and statistics to highlight how the brand's products benefit overall health by reducing contaminants in drinking water turned users just window-shopping on the homepage to engaged customers.
>
> This form of social proof encouraged product page visits, leading to $410,000 in annualized revenue gains.

Multimedia Content

Your company's photos and videos play an important role in framing the products on their screens. Remember, customers can't physically see or touch these products. If it's an app or service, your offering is even less tangible. Media content visually fills those gaps, leaving the customer with a more complete impression.

Let's say your company wants to frame itself as a high-quality solution. The price may reflect the quality. Reviews may rave about the product, and it could be featured in any number of gift guides or "best-of" rankings. But that's only if the customer does that extra bit of digging. If their first impression is simply a photo that doesn't say much about the product itself—or worse, is low quality—they're likely not going to think about digging deeper.

INFORMATION-GATHERING

But your company can frame your product to fit that message through visual cues, such as:

- **Usage:** Before you dive into the subtle details that convey a message about your product or service, you need to make sure you have the basics covered first. Product demos—such as a GIF or video—give a straightforward and clear explanation of the item itself.

    None of the following tactics will be of any use if the customer has no idea what the product or service is and how it works.

- **Setting:** Where you showcase your product will tell your customers everything they need to know about it.

    If a company wants to position themselves as the best, the most high-tech and sophisticated laptop for college students, they may use photos of the computer on an ivy-strewn campus, not sitting on an apartment coffee table next to an empty Easy Mac container. Both images say "college student," but one is certainly more elevated than the other.

    But on the other hand, if the company wants to position itself as a more relatable company with a laptop that gets the job done and is with you through the late-night study sessions and mad dashes from the library to your next class, that coffee table image may be the better choice.

- **Association**: Companies can signify a quality about their product by paring it with another product that's well known for that same thing.

    For instance, when Bang & Olufsen headphones are displayed next to a $300 stovetop espresso maker, it automatically elevates the perception of their quality.

- **Endorsements**: Another way companies can borrow credibility is by associating themselves with other high-end companies or figures.

    A perfect example is Nespresso's collaboration with George Clooney. Clooney isn't just a famous actor; he's a symbol of luxury and elegance, with properties like a house on Lake Como. Associating their company with him lends Nespresso an air of prestige.

But what happens if a company frames their products for the wrong audience? You'll remember in a previous chapter, we talked about a baseball equipment company. They made the mistake of creating flashy videos aimed at kids. These videos featured a cool, ex-college baseball player who was loud and in your face, catering to a younger YouTube audience.

However, the parents, who were the actual consumers, just wanted to know if the bat was right for their child. They would often lose interest and not watch the video to completion. Framing the product in the wrong way alienated their true audience, which, in turn:

- Confused the customer by introducing irrelevant information
- Disrupted and slowed their digital journey, making it harder to ultimately convert
- Wasted valuable space, time, and resources by creating and hosting a video that did more harm than good

A Word of Caution

Framing can be a powerful tool in conveying a positive message, but you must tread carefully. Misleading framing strategies can lead to higher return rates and a loss of consumer trust.

- **What framing is:** Presenting information in a factual and intentional way that helps customers navigate their choices and arrive at the best decision for their needs.

- **What framing is not:** Manipulating or cherry-picking facts to give customers a falsely positive perception of the company or product or tricking them into buying something that doesn't solve their problem or need.

Being aware of the ethical implications is crucial—you don't want to cross the line into manipulation or deceit.

Framing should be deployed with particular care with audiences that are more knowledgeable about the product. A customer ordering a sponge is going to glance at the information they need and call it a day. A customer browsing specialty fair-trade, organic coffee beans because they can't stand the idea of going to Starbucks is going to be able to read through the fluff.

Let's See This in Action

A few years ago, Skechers Shape Ups landed itself in hot water for misleading customers about the shoe's benefits.

In 2013, Skechers' launched an infamous ad campaign with Kim Kardashian, claiming that she didn't have to go to the gym because her Sketcher Shape Ups provided enough of a workout. It's likely that the company wanted to frame the shoes as a great footwear option for those serious about getting into shape.

> The Federal Trade Commision accused the company of misleading customers "by making unfounded claims that Shape-ups would help people lose weight, and strengthen and tone their buttocks, legs and abdominal muscles."[44] In fact, many customers who bought the shoe actually gained weight because they stopped going to the gym, relying solely on the shoe for exercise.
>
> So an ad campaign meant to keep Skechers' sales flying off the shelves instead brought a $40 million class-action lawsuit.
>
> It goes to show that while framing can be powerful, it also comes with responsibilities and potential repercussions.

Putting It All Together

Everything your company has on its website is framed in some way, whether it's intentional or not. The digital journey offers opportunity to frame your product or service in a certain way through:

---

[44] https://www.ftc.gov/news-events/news/press-releases/2012/05/skechers-will-pay-40-million-settle-ftc-charges-it-deceived-consumers-ads-toning-shoes

- Product or service descriptions
- Social proof
- Multimedia content

But remember: framing and lying are not the same thing. Tread carefully.

SOLUTION 2:
SAY IT AGAIN...AND AGAIN...AND AGAIN

Just because the information is on the website doesn't mean your customers see it.

Imagine you're in a crowded hallway, waiting to enter a theater to see the nine o'clock show. You look over your shoulder and notice a man who stands at least a foot taller than anyone else in the room. He even has to duck his head to get through some of the archways.

You may interact with dozens of people over the course of the evening—waiters, other patrons, an Uber driver, etc.—but thanks to something called salience bias, you'll probably remember way more details about that man than anyone else.[45]

---

45 Armenia, G. (2013). Lazy Thinking: How Cognitive Easing Affects the Decision Making Process of Business Professionals. *Honors College Theses*. 126. https://digitalcommons.pace.edu/honorscollege_theses/126.

That's because we tend to direct much more of our focus on things that are noteworthy and ignore things that aren't attention-grabbing. That doesn't change when we sit at our computers to shop.

There are a few psychological reasons this happens:

- We're more likely to notice certain things over others.
- We don't remember all the information we encounter.

Reason 1: We're More Likely to Notice Certain Things over Others

Imagine you're at a supermarket. You enter an aisle filled with rows of various cereal boxes. Among them, there's one box with incredibly colorful and eye-catching packaging. It has bold graphics, glittering effects, and vibrant colors that make it stand out from the others. This box is promoting a new, limited-edition cereal with a catchy name.

Now, within the same aisle, there are several other options that are less visually striking. They have plain, traditional packaging and are known for their nutritional value and health benefits. These cereals might have a simple, unassuming design. If you compare them side by side, you'll see that they are the same exact cereal, just from different companies. But guess which one often makes it into the cart.

That is salience bias at play in our consumer habits: people naturally focus on items that are visually appealing and overlook those that are less eye-catching. This typically comes from some type of contrast between an item and its surroundings. It's like suddenly hearing someone's cell phone ring in the middle of a play—the sudden noise in an otherwise silent audience is jarring and not something you're likely to miss.

There's a good reason this happens. Salience bias comes from the core of our evolution—to survive, people needed to be on alert when something seemed off, whether it was listening for the arrival of a predator or knowing not to eat the bright pink moss grossing on the side of the tree because it may be poisonous.

Salience bias isn't as life-or-death today, but it exists nonetheless. The faster paced the world becomes, the more valuable your customers' attention is. They don't give it out for free—you have to earn it. And if you want to turn that attention into conversions, you need to make sure that you're drawing attention to exactly what your customers need to make the best decision.

Let's talk about how.

Featured Product or Service Placement

Salience bias can manifest when a prominently displayed "featured product or service" catches the shopper's attention.

This featured item might not necessarily be the best fit for the shopper's needs, but its eye-catching placement and design make it stand out in contrast to the surrounding product listings. Shoppers might be drawn to this product despite other options being more suitable.

> ### Let's See This in Action
>
> An attention-grabbing element isn't always a good thing.
>
> We saw that play out with one of our clients at The Good while testing featured products.
>
> During testing, session recordings and heat maps showed early customer drop-off on desktop and mobile product pages but heavy engagement with page elements such as recommended products and featured categories.
>
> Based on that behavior, we showcased products and categories toward the top of the homepage to clearly show navigation elements above the digital fold.
>
> This change increased engagement—but it lowered conversions.
>
> User behavior indicated that this test should have worked, so we had to look further. Taking a step back in the customer journey, we saw that the majority of

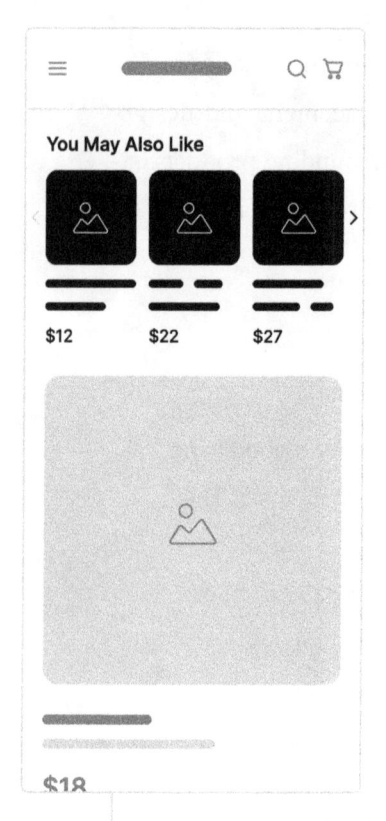

users came in from Facebook ads straight to a product that was very well targeted to their interests.

In other words, customers who were most likely to buy were already where they needed to be.

By adding navigation elements that grabbed attention away from the product page, we inadvertently encouraged further browsing when none was needed, making them less likely to check out.

The initial decision was easy—it was what brought them to the site in the first place. By adding appealing additional alternatives, we both distracted the customer from converting and encouraged them to dive even deeper into a website they likely hadn't been to before. This caused analysis paralysis, effectively launching them back from the Decision-Making Phase to the Information-Gathering Phase.

INFORMATION-GATHERING

Flashy Banner Ads

Online retailers often use banner ads with bright colors, animations, or attention-grabbing text to promote products, new services, or special offers. These visually striking ads can trigger the salience bias, diverting shoppers' attention from the main content of the webpage. Shoppers may click on these ads impulsively, even if the advertised products aren't what they originally came to the website for.

This can work for companies or against them. If you catch a customer's attention with a banner ad, then you need to make sure that you fulfill whatever promise you made that prompted them to click in the first place. All too often, sites use a banner that highlights one of their bestselling products, but when customers click for more information, they're brought straight to the "shop all" product or service page.

Now, the customer is frustrated and jolted off their purchasing path—and it's going to hurt conversions.

Let's See This in Action

We encountered exactly that with one of our clients at The Good. The problem: the homepage hero received a lot of clicks. But conversions from there? Not so much.

Using session recordings and user testing, we found that people who clicked on the homepage photo expected to be taken straight to that product page. Instead, they were directed to a general "shop now."

Users didn't end up where they expected to, which not only interrupted their customer experience but also took away some confidence in a purchase they were very intent on before

So how did we fix it? Simple: we gave the people what they wanted. First, we changed "shop now" to "shop all" so users expected to browse. Under that, we added an "explore product" button so anyone interested in the highlighted product could get there easily and efficiently.

That simple change increased conversion rate by 75 percent.

It may seem straightforward, but when you know your website backward and forward, you might forget that other people don't. If they can't figure out how to intuitively navigate your website and your products, they're much less likely to convert.

INFORMATION-GATHERING

Benefit Squares

Introducing callouts to a product or service page makes it clear what's most important. Instead of bombarding customers with every bit of information and data, you can distill it into one statement. Then, implement it in an eye-catching way as benefit squares on your product pages.

Let's See This in Action

This tactic brought In more than $684,000 in annualized revenue for one of our clients.

While conducting user testing, click maps showed that many customers turned their attention to carousel images. This told us that they really spent time analyzing products before making a purchase decision.

Careful consideration is great. But the harder the customer has to work to make a decision, the less likely they are to come to one at all.

Implementing "Best Seller" and "New" product flags gave

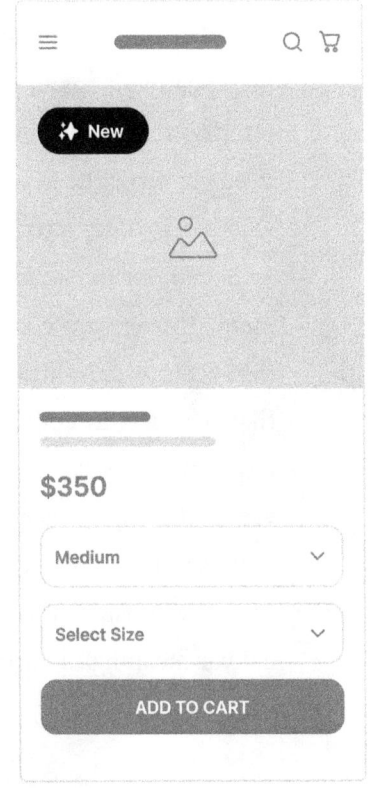

> customers a shortcut toward narrowing down their search—which meant more customers heading to check out.

Headlines

Headlines are the biggest text on the page, which means they grab more attention than anything else. Do not want to waste precious headline space saying nothing.

Many people opt to use headlines as a lead-in or a question, when, in reality, the headline should aim to answer a question. The customer already knows what their question is, and by repeating it again, you waste space designed to catch their attention. Ideally, readers should be able to glean what they need just from the headlines, which can then help them decide whether to delve into the body copy.

Let's consider a SaaS product related to project management. If users are concerned about the software's collaboration features, the headline should straightforwardly convey that it excels in collaboration. If the headline merely poses a question like "Is it great for collaboration?" or provides a vague statement such as "Exploring our collaboration capabilities," it falls short in efficiently serving the reader. This not only consumes their time unnecessarily but also fails to deliver meaningful information right from the start.

INFORMATION-GATHERING

Or in e-commerce, let's say you're concerned about a product being vegan or plant-based—such as a food creamer. The headline should straightforwardly tell you it's plant-based. Then, the body copy could elaborate on the benefits or certifications. A headline that simply poses a question like "Is it plant-based?" or vaguely states "How we approach..." doesn't serve the reader well. It wastes their time and provides no meaningful information.

Let's See This in Action

Headlines draw customers in. But once you grab them, you'd better deliver what they promised.

For a client at The Good, session recordings showed that customers were quick to click on the features page, but the content didn't necessarily match their expectations.

And there's nothing more annoying than that.

By updating the layout of the features page, we were able to effectively hone in on the information customers were really looking for—which, in this case, was the product benefits.

Highlighting language and pictures of the products that matched the headline that brought the user there to begin with brought a 12.23 percent increase in conversions.

Putting It All Together

If you want your customer to pay attention to something specific, you have to make them pay attention. You can do this through:

- Featured product or service placement
- Flashy banner ads
- Benefit squares
- Headlines

Reason 2: We Don't Remember All the Information We Encounter

Understanding and accounting for salience bias holds extra importance when you realize that even when you do manage to grab the customer's attention, it doesn't mean they're going to remember it.

This is called the Google effect, which suggests that people tend to forget information that they can readily find in search engines like Google.[46] It has nothing to do with their memory itself. Because the brain stores so much information throughout

---

46 Rowlands, I., Nicholas, D., Williams, P., Huntington, P., Fieldhouse, M., Gunter, B., Withey, R., Jamali, H. R., Dobrowolski, T., & Tenopir, C. (2008). The Google generation: The information behaviour of the researcher of the future. *Aslib Proceedings*, 60(4), 290-310. https://doi.org/10.1108/00012530810887953

the day, it creates shortcuts to maximize how much information it can take in. There's no need to store something in your memory if Google's already storing it for you.

Imagine you're planning a weekend getaway and come across a fantastic restaurant recommendation on a travel website. You think, "I'll remember this for sure; it's a great find!" Then, when the weekend arrives, you find yourself in the destination, hungry and eager to try that amazing restaurant. Except...you can't recall its name or location because you didn't bother committing it to memory. After all, you thought Google would have your back.

This happens again and again in every aspect of your life. Studies have even shown that people who take a lot of photos and videos of a particular event won't remember it as well as someone who kept their phone in their pocket.[47] The photos and videos in their camera roll mean they can relive the moment at any time, so there's no need to commit it to memory. Never mind the fact that no one has actually ever gone back into their photo album to watch videos of the firework show at last year's Fourth of July party.

This digital amnesia means that it's not enough to put information on your website once and assume the customer saw it.

---

47   Henkel, L. A. (2014). Point-and-Shoot Memories: The Influence of Taking Photos on Memory for a Museum Tour. *Psychological Science*, 25(2), 396-402. https://doi.org/10.1177/0956797613504438

INFORMATION-GATHERING

If it's important enough to highlight, it needs to be included in every area where it's relevant. Just as that restaurant recommendation would have been better off written down, important information on your website should be repeated for your customers to ensure they remember and act upon it.

This affects a number of website elements.

FAQ Pages

Every company thinks they need an FAQ page. It covers all your bases, right? Any questions the company thinks a customer might have are all neatly wrapped up into one page. While that seems like a great way to ensure customers always have access to the information they need, it's almost always a bad idea.[48]

Many (maybe most) digital managers falsely believe that FAQ pages rank right alongside the about page and contact page as necessary components of a professionally designed e-commerce website. The problem, though, is that FAQ pages tend to become the dumping ground for sloppy content, lazy SEO, and poor customer insight. Instead of offering a space for customers to find the information they need all in one place, it gives a false sense of security that everything the customer needs to know is answered in the buying process.

---

48   https://thegood.com/insights/faq-pages

Just because it's answered on the website doesn't mean it's answered in the buying process. In fact, the need to have an FAQ page at all is a clear signifier that your customers aren't getting the information they need.

Following this heuristic, simply having an FAQ page dooms every piece of information on the website right off the bat. Customers know exactly where to find information if they need it, and they may gloss over everything else on the digital journey without realizing. The problem is, just because they know where to find it doesn't mean that they're even going to make the effort to seek it out. And even if they do, it takes the user out of the digital journey into a completely different part of the website—one that doesn't contribute to the purchasing funnel. From there it's just a gamble of whether they make their way back.

Let's say, for example, Mike is looking for a new fishing pole. He's concerned about the quality of the materials, but the product page doesn't tell him much. He's already in a position to buy, but now he has to navigate away from a commercial page to visit your FAQs. Will Mike add a product to the cart? We can't say for sure, but the odds are a lot lower than if the information was presented to him organically where he expected it.

Adding the relevant information along the digital journey—as opposed to hiding it in the FAQ—is especially important for SaaS offerings. It's important to ensure that customers feel like they already grasp the software or app's workings and features right from the start. SaaS products or services frequently introduce users to a whole new realm of functionality, and it's essential to present information in a manner that fosters an immediate sense of understanding, even for those who haven't yet experienced the software or app firsthand. Even a momentary lapse in clarity can potentially lead customers to consider other, more straightforward alternatives.

Instead, it is better to have all of that content directed at your product or service pages—the pages that generate revenue.

Let's See This in Action

With this in mind, we raised a red flag immediately when we saw heavy customer engagement on one of our clients' FAQ pages.

This showed us that customers were having a hard time finding the information they needed to make a decision on the product page alone. The product in this case was courses, so to address this without taking the customer off of their digital journey, we incorporated a course overview directly on the product page.

> [mockup of course product page showing $1,500 price, ENROLL button, star rating, and "What This Course Has to Offer:" section with four icons: 99th Percentile Instructors, 100 Expert Strategies, 3 Days a Week, 4,500+ Verified Reviews]
>
> There, we answered all relevant questions quickly, succinctly, and before the customer even thought to ask them, which increased confidence in their enrollment decision and resulted in a $773,000 annualized revenue boost.

Product Pages

In a similar vein to the FAQ page, when it comes to getting your customers to remember the information they need to make a decision, your product page can either further reinforce every message they've received along the way, or completely disregard it.

INFORMATION-GATHERING

In earlier stages of a healthy discovery and information-gathering journey, your users should have some idea of the main products, services, and values your company offers. But just because they saw it once doesn't mean they're going to carry that over when they hit the product page—either because they didn't make the connection the marketing team felt was so obvious or simply because they forgot.

It's like when you're walking through your home to get something from the kitchen. You get up from the couch, pass through a doorway from the living room into the kitchen, and stop in your tracks because you suddenly have absolutely no idea what you came in there to do in the first place. This is called the doorway effect. When we change physical or mental environments, we tend to forget things along the way.[49]

The same is true for any type of digital path. Every new page a customer has to navigate to offers a higher chance for the information from the previous page to slip away. This is why creating a digital journey that makes decisions as quick and easy as possible is the best way to boost conversions.

Which is all to say that just because you said it earlier doesn't mean you can't say it again. If a piece of information

---

[49] https://www.bbc.com/future/article/20160307-why-does-walking-through-doorways-make-us-forget#:~:text=The%20Doorway%20Effect%20occurs%20because,forgotten%20when%20the%20context%20changes.

is so vital to the decision-making process that the company is willing to lead with it, then it belongs in every space where it is relevant.

Imagine you're an e-commerce business selling smartphones, and one of your key selling points across all your product lines is that they come with a free protective case. Now, instead of assuming that customers will remember this detail from your homepage or a separate promotions page, it's crucial to repeat this valuable information on every individual product page. By ensuring that the inclusion of a free protective case is reiterated on each product page, you maximize the chances of customers considering this valuable offer in their purchasing decision, ultimately boosting conversions.

The key here is to address the customer's question before they have a chance to ask it—even if you feel like you've answered it already. Actually, especially when you feel like you've answered it already.

Let's See This in Action

There's a reason that we feature free shipping language on every client page we can get our hands on—banner, buy box, cart, checkout, product page—and that's because if it's that important, it's worth repeating.

INFORMATION-GATHERING

For this client, we saw heavy user engagement in the buy box area, which told us that customers were engaged in learning more about the product.

That also meant that the product description didn't answer the questions it needed to in order for them to make a decision.

Updating the product description to better speak to user questions motivated users to purchase and created another place to access this key information. And it increased conversions by 4.73 percent.

Many people assume that users will eventually see some messaging on your website, but most don't go to all your pages and pore over everything before making a decision.

So you have to cover your bases if you want to make sure your top selling points are front and center no matter what stage of the digital experience your customer is in.

> When it comes to information that drives a decision, repetition is key.

Search Bar

The Google effect not only affects what customers remember from page to page, but how they go about finding that information at the moment they realize they need it. Because customers subconsciously know that they can search for any additional information they need, it's crucial that companies have a functioning search bar. The problem is, simply having a search bar isn't enough.

Many website managers assume that search is only used by customers who know exactly what they're looking for—and in a lot of cases, they do. But not all.

Baymard Institute ran a study based on site search usability and found that shoppers tended to fall into one of four search categories:

- **Exact search**: The user knows exactly what they're looking for right down to the exact name of the product (e.g., "iPhone 13").

- **Product type search**: A user searches for a general product rather than a specific company or model (e.g., "laptop").

INFORMATION-GATHERING

- **Problem-based search:** A user searches for a solution to a problem without knowing exactly what kind of product they need (e.g., "acne remedy").

- **Non-product search:** The user is looking for information rather than a product (e.g., "opening times" or "shipping options").[50]

It is the last two types of searchers that really benefit from— or become stuck by— the Google effect.

These are the customers who think they can find what they need without retaining the information along the way. And if they go to find it and they can't, it wipes out all the information they've gathered up until this point, frustrates them, and ultimately is likely to make them abandon the website completely.

> ### Let's See This in Action
>
> Sometimes, customers just need a little nudge to keep them on the right track. Check the most common search terms on your website—if they often include vague descriptions instead of specific queries, you may want to consider running an instructional search test.

[50] https://baymard.com/research/ecommerce-search

Instructional search encourages intentional browsing, improves user experience, and boosts conversions. Testing a friendly, instructional search prompt may be exactly what you need.

During research for one of our clients, user tests and session recordings revealed that customers primarily navigated through the search bar. We also found that customers only engaged with select menu categories.

Our team hypothesized that adding friendly microcopy and enhancing search bar visibility would encourage the use of search and, in turn, increase transactions. We visually emphasized the search bar with a white background and updated the language to "Try 'search term.'"

It delivered over $3 million in revenue gains.

Why does it work? Fifty-nine percent of web visitors frequently use a website's internal search navigation, and 15 percent would rather use the search function than the hierarchical menu.[51] You don't want your

---

51   https://ux.stackexchange.com/questions/18089/how-important-is-the-search-box

customers to zombie scroll through endless products to find what they need. Internal search encourages customers to easily locate what they need.

If you want to run this test on your website, you could try:

- Adding friendly microcopy to your search bar (include a key product or category for inspiration)
- Exposing the search bar on mobile
- Adding a white or light background to the search bar

Putting It All Together

Anything worth saying bears repeating—especially if it's going to help make the sale. Because even if you say it once, customers may not see it. Even if they see it, they may not remember.

Here are a few elements this affects:

- FAQ pages
- Product pages
- Search bar

QUESTION:
WHO ELSE HAS ALREADY BOUGHT THIS, AND IF I JOIN THEM, WHAT DOES THAT SAY ABOUT ME?

As much as we love to think that we all stand out from the rest in some way, it's human nature to want to fit in. Again, it goes back to evolution—the bigger and stronger the group you lived in, the safer you were from outside predators. Today, it's more like the more Instagram likes you have, the more popular you are.

No, we know it's way more deeply rooted and nuanced than that. But the point stands—human beings are social creatures who want to be accepted and admired. We don't want to look foolish or make mistakes. So we often subconsciously do what others are doing, even if we pride ourselves on individuality,

To do that, we follow what are called social norms. These are the commonly accepted beliefs of what behavior is—and isn't—appropriate for a given situation. These norms vary from culture to culture, place to place, and era to era. But no matter what the actual norms are, they still exist in some form or another, and they heavily influence the way people make decisions.[52]

---

52   McLeod, S. (2016). What is conformity? *Simply Psychology*. https://www.simplypsychology.org/conformity.html#sherif

In many ways, these social norms that guide interactions in the physical world can also be applied in the digital realm. When making a purchase, a user's main goal is to solve a particular problem. But underneath that are so many underlying questions that help them get to that point. In this context, the big question is, "What does this purchase say about me?"

SOLUTION:
SHOW THEM EXACTLY WHAT THIS PURCHASING CHOICE SAYS ABOUT THEM

The best way to show customers who they'll be if they buy this product? Show them who else has bought it.

This is where social proof comes back into play, just in a different way. When visitors shop at your store, they look for clues from trusted sources to help them make decisions. They might consult reviews, testimonials, or expert recommendations. They might watch videos from influencers they trust or solicit recommendations from organizations, nonprofits, or other entities.[53] Ninety-three percent of consumers say that online reviews influence their purchase decisions.[54]

Generally speaking, there are seven types of social proof:

---

53 https://thegood.com/insights/social-proof
54 https://www.qualtrics.com/blog/online-review-stats

- **Expert:** A trusted authority in an industry or niche recommends a product.

- **Celebrity:** A celebrity endorses a product. It's effective even if we know the celebrity was paid for the endorsement.

- **User:** A user or customer recommends your products based on their personal experiences (e.g., ratings, reviews, social media comments, etc.).

- **Wisdom of the crowd:** A large group of people endorses your company (e.g., millions of social media followers).

- **Wisdom of your friends:** A customer's friend recommends a product. We tend to weigh recommendations from our circle of friends more heavily.

- **Certification:** A product or company receives a stamp of approval from a trusted authority.

- **Earned media:** The press publishes positive stories about your company.

INFORMATION-GATHERING

Leveraging these types of social proof can give your customers the information they need to make a decision they feel confident in. Here's why.

People Care What Others Think about Them

Think back to the last time something embarrassing happened to you in public—you just knew that everyone was looking at you, even though you probably can't remember the last time something embarrassing happened to someone else. And even if you can, you probably don't think it was that big of a deal (even though if it had happened to you, your whole world would have come crashing down).

That feeling that people are watching you, even though you're not watching others, is called the spotlight effect. It manifests in various aspects of our lives, making us overestimate our significance, both in positive situations, like when we excel at something and think everyone is highly impressed, and in negative situations, like when we fail and assume that everyone is ridiculing us behind our backs.[55] This is a psychological phenomenon where individuals tend to believe that they are the center of attention and that others are closely scrutinizing their actions. Even their digital purchases.

This perceptual bias often intensifies the perceived significance of a consumer's choice, leading them to believe that their purchase will be under the spotlight, subject to scrutiny by peers, family, or even strangers. This heightened self-awareness can have significant implications for their decision-making process, causing them to opt for products or services that closely align with societal expectations or their desired self-image, sometimes at the expense of their actual needs or preferences.

---

55 Gilovich, T., Medvec, V. H., & Savitsky, K. (2000). The spotlight effect in social judgment: An egocentric bias in estimates of the salience of one's own actions and appearance. *Journal of Personality and Social Psychology*, 78(2), 211-222. https://doi.org/10.1037/0022-3514.78.2.211

INFORMATION-GATHERING

People Want to Fit In

Have you ever seen a celebrity wear a certain item or seen a TikTok video using a particular product go viral, and all of a sudden everyone around you has that product? This is thanks to the bandwagon effect.

The bandwagon effect is a psychological phenomenon where people tend to do something primarily because others are doing it. The influence can be global—like an influencer endorsement—or local—like when you go to the dog park and see a lot of the families have the same water bottle for their pup, and suddenly you're online buying it too. Especially because we now have the ability to whip out a phone and buy just about anything we see someone else has, at any time, from anywhere, the bandwagon effect has a significant influence on purchasing decisions.

The key here is to understand that this is not about exploiting people's desire to fit in—it's about being clear about how your product helps users connect to the people around them in the way they want to.

A great way to take advantage of the bandwagon effect is through customer reviews and ratings. Showcasing customer reviews and star ratings provides immediate social proof.

- **Reviews:** As shoppers, we tend to consider reviews as more trustworthy than the copy written by the seller. After all, in most cases, the reviewers are giving their honest opinions about the product without any ulterior motives.[56] Ninety-seven percent of consumers read product reviews and ratings before making a purchase decision.[57] When downloading an app, for example, reviews provide potential users with valuable insights into the app's functionality, reliability, and user experience. Positive reviews can instill trust and confidence, while negative reviews may deter users from

---

56 https://thegood.com/insights/product-reviews-improve-conversion-rates
57 https://www.powerreviews.com/insights/2021-ugc-conversion-impact-analysis

downloading at all. No one wants to use up valuable space on their device for something other users tell them isn't worth it.

- **"Bestseller" or "Most Popular" section:** These sections inherently suggest that many others have made the same choice and joining in will make you part of this popular group.

- **Wishlist:** Displaying the number of people who have added an item to their wishlist can imply popularity and prompt new customers to consider doing the same.

- **Real-time notifications:** Some websites display real-time notifications of recent purchases by other customers, often accompanied by their location. This can be especially powerful for high-consideration products or services, making potential buyers feel they're making a wise, well-vetted decision.

- **Waiting lists:** When you run out of stock, this can present an opportunity to show your customers just how popular that product is. Seeing that there's a waiting list and a mention of upcoming batches creates a sense of urgency and desire—without playing into scarcity scare tactics like "only 10 left." When potential buyers perceive that a product is in high demand and that others are eager to get their hands on it, it often sparks a desire to join the queue, as they believe that it must be something valuable.

But be careful—while you can benefit from appealing to a large number of customers, casting your net too wide can hurt more than it helps. Trying to appeal to a vast and diverse audience makes it challenging to conduct effective market research because the focus is spread thin.

People Want to Be a Part of Something Bigger than They Are

Often, purchases signal something much bigger about the purchaser than the solution they're trying to solve.

If someone purchases a new pair of hiking boots, it often doesn't *just* mean that they like to hike. It means that they're a part of a community of like-minded people—people who carry thermoses of coffee for miles in the dark just so they can have their first sip of the day on a mountain peak watching the sun rise.

In the realm of SaaS, consider the example of Canva versus Adobe Photoshop for graphic designers. Both are exceptional products, offering users the tools to create stunning designs.

INFORMATION-GATHERING

However, the distinction lies in the user base they tend to attract. Canva is often seen as a user-friendly design tool catering to a broader audience looking to create eye-catching visuals quickly. On the other hand, Adobe Photoshop is renowned as a professional-grade software favored by experienced designers, illustrators, and photographers for its advanced capabilities. The service itself says something about the type of professional you're hiring.

The community your product or service impacts isn't something you should leave to interpretation. It's up to the company to clearly show the customer that if they buy this, this is the type of person they're going to be. The more clearly you convey that message, the more likely the product is to convert.

This is influenced by a phenomenon called in-group bias. It's not only the desire to be part of a group, but the idea that people have a tendency to gravitate toward products that clearly state something about their identity and the larger community that comes with it.[58]

This phenomenon manifests in big ways and small, on purpose and by accident. In fact, group membership doesn't need to be based on anything particularly meaningful for the bias to emerge—so long as it resonates in some way, it's enough.

But while you can play on elements that highlight your products' in-group benefits, you can't force it. Used correctly, in-group bias isn't you telling customers who they are or what they care about—it's showing them that you have a deep understanding of the people who are already buying and reflecting that community back through:

- **Images:** Imagery is one of the most powerful ways to help customers feel more personally connected to your products. It's not just about showcasing your items—it's about creating an environment where customers see themselves seamlessly fitting into your company's narrative.

---

58 Cadsby, C. B., Du, N., & Song, F. (2016). In-group favoritism and moral decision-making. *Journal of Economic Behavior & Organization*, 128, 59-71. https://doi.org/10.1016/j.jebo.2016.05.008

INFORMATION-GATHERING

Take, for instance, businesses like Good American, renowned for prominently featuring models of all sizes, or Savage by Rihanna, celebrated for their commitment to showcasing a diverse range of individuals. These companies serve as prime examples of how strategic imagery can instill a profound sense of belonging within customers. This visual storytelling, which aligns with values of inclusivity and representation, fosters an in-group bias by validating and celebrating the diverse identities and backgrounds of potential customers. As a result, customers not only feel seen but also feel an authentic connection to the company, strengthening their loyalty and trust.

- **Copy:** Consider what's important in your customer's decision-making process, and make sure it's reflected in the copy in a way that adds a personal or human element. This will mean something different for every company.

  For a company that prides itself on being a part of the community, it's going to be reassuring for customers to see that your headquarters are based in their city or that you're actively pursuing initiatives to give back to your community. If the majority of your customers are women, it may be reassuring to know that the company is owned and run by women. The core principle here revolves around direct engagement with a specific audience. When crafting your message, avoid making generic, sweeping statements that could be applicable to anyone. Instead, tailor your communication to speak directly to the concerns, values, and interests of your target demographic. By doing so, you avoid the trap of creating content that is aimed at everyone but resonates with no one.

- **Use cases:** There's no need to be subtle or coy along the digital experience. When you clearly communicate what your product or service does and how to leverage its features, customers perceive it as tailor-made for their specific needs.

    For instance, a management platform needs to highlight how its features cater to various scenarios. They could emphasize that their software is exceptionally suited for marketing teams seeking streamlined campaign management or for IT departments in need of robust issue tracking. While the platform may serve both purposes, by addressing these specific use cases, the company can forge stronger connections with their audience and simplify the decision-making process for potential users.

INFORMATION-GATHERING

Human nature drives us to seek acceptance, and this extends to our choices as consumers. We want to know who else has made the same decision and what it says about us. Social proof, encompassing various forms such as expert endorsements, user reviews, and the wisdom of friends, plays a pivotal role in guiding these decisions.

Understanding and leveraging these psychological principles throughout the digital experience can significantly enhance the digital journey, boost revenue, and strengthen brand loyalty.

Putting It All Together

Social proof has a profound effect on people's purchasing decisions—whether they'd like to admit it or not. There are a few reasons for this:

- People care what others think about them.
- People want to fit in.
- People want to be a part of something bigger than they are.

The better you address these fundamental social needs, the more likely they are to choose you above your competition.

AT A GLANCE

In the Information-Gathering Phase, Your Customer Will Ask

- Can I find what I'm looking for?
  - **Solution:** Create a clear and customizable browsing experience

- What do I need to know about this purchase?
  - **Solution 1:** Leave nothing up to interpretation
  - **Solution 2:** Say it again...and again...and again.

- Who else has already bought this, and if I join them, what does that say about me?
  - **Solution:** Show them exactly what this purchasing choice says about them

Principles at Play

- **Maximizers:** Customers who need to know every available option in order to make the best decision possible.

- **Satisficers:** Customers who are content to find a good solution to their problem and don't need to continue their search.

INFORMATION-GATHERING

- **Serial positioning effect:** People best remember the first and last items in a series and find it hard to remember the middle items.

- **Primacy effect:** The tendency to better remember information at the beginning of a series.

- **Recency effect:** The tendency to better remember information that was most recently told to us.

- **Choice overload:** The tendency for people to become overwhelmed when faced with too many options.

- **Framing effect:** Our decisions are influenced by the way information is presented.

- **Herd mentality:** We presume that other people have more knowledge about the situation, so we should follow their lead.

- **Salience bias:** People naturally focus on items that are visually appealing and overlook those that are less eye-catching.

- **Google effect:** People tend to forget information that they can readily find in search engines like Google.

- **Doorway effect:** When we change physical or mental environments, we tend to forget things along the way.

- **Social norms:** The commonly accepted beliefs of what behavior is—and isn't—appropriate for a given situation.

- **Spotlight effect:** The idea that people are watching you and judging your behavior more closely than they are.

- **Bandwagon effect:** People tend to do something primarily because others are doing it.

- **In-group bias:** People have a tendency to gravitate toward products that clearly state something about their identity and the larger community that comes with it.

Optimization Opportunities

- Filters
- Category pages
- Navigation menu
- Search bar
- Product or service descriptions

INFORMATION-GATHERING

- Social proof
- Multimedia content
- Featured product or service placement
- Benefit squares
- Headlines
- FAQ page
- Search bar

In Summary

- How the information is presented matters more than what the information is.

- If you want customers to find the product or service they're looking for, you need to streamline the search process across any element that helps navigate the digital journey.

- Information presented should be clear and concise and should directly address the questions that shoppers actually have.

- Customers don't need all the details—they need the right details. Distill the initial information customers see to the most essential details and then give an option to seek more information if they need it.

- Facts aren't enough—you have to present the facts in a way that frame the message you're trying to convey (while staying truthful).

- Framing must be used ethically. Misleading framing strategies can lead to higher return rates and a loss of consumer trust.

- If you want your customer's attention, you have to either draw attention to the element you want them to see or maximize the value of the elements they are naturally drawn to.

- Customers are likely to forget information as they go through the digital journey—don't be afraid to remind them along the way.

Your careful guidance along a curated digital journey hits its most pivotal moment when it becomes time to make a decision and convert. The ease with which customers move from information collection to decision-making, and ultimately to checkout or sign-up, presents a unique chance to make this decision seem like an inevitable end to their journey. The more straightforward and guided the process is, the more easily customers can decide and convert.

4

DECISION-MAKING AND CONVERSIONS

Once your customers gather all of the information they need, it's time for them to make a decision.

Companies that insist on visitors browsing their entire digital experience to "fall in love" with the company are missing the point. Most of the time, visitors have a specific purpose when shopping online. Maybe they're comparing a few products across different websites, but it's all aligned with the same "job to be done." They aren't aimlessly wandering through digital aisles.

And while you've been guiding them along the way up until now, the decision-making and conversion process is where you really take the lead. You don't want customers to linger—you want them to convert.

This is where we once again step back to talk about the difference between perception and reality. When we draw out every step of the customer's digital journey and every question along the way, it gives the impression that each user is carefully vetting every piece of information about your products and their decision.

Yes, some customers do carefully vet each decision. Someone investing in a new, expensive couch they know they'll keep no matter where they move next is likely going to invest time and effort to make sure it's the best choice for them. But for many other products or services, customers just want to make a decision that solves the problem or need at hand.

What's taking you hours to carefully think through as you read this book may take your users only a fraction of a second to experience. But whether that Information-Gathering Phase lasted a few hours or just a few minutes, customers experience that same cognitive load. The difference isn't in what customers need to consider; it's how well you guide their path to making these decisions.

This is where many companies start to falter—they may assume the decision to buy a product like soap is pretty unimportant, so it should be a quick, no-fuss situation. But by not giving customers all the information *just in case* they need more, they make their product or service just as easy to overlook as it is to add to cart.

Others draw out every piece of information too much, overwhelming the customers and ultimately burying the information they need to make a decision in fluff and extra details that are useful *if* they're needed but they aren't always needed.

The key is to approach every product the same way—by asking yourself, "How do we get them from their first touch point with our company to conversion as efficiently as possible?" It looks different for each customer and business, but there are plenty of ways to optimize. Then, execute accordingly. The goal is to make the right decision for them as much of a no-brainer as possible.

To help customers along and get out of their way in the process, you have to understand what considerations are involved in getting them from "Should I purchase?" to "Check out."

Customers may ask themselves:

- Am I ready to make a purchase?
- What should I choose?
- If I'm wrong about this choice, what is the worst possible outcome I could experience?
- What's in it for me?
- How easy is it for me to purchase?

QUESTION:
AM I READY TO MAKE A PURCHASE?

The decision, at first, isn't *what* they want to buy. The decision is whether they're ready to buy anything at all.

If they aren't ready by now but you know your products are perfect for your target market and your marketing team brings that target market to the website, that means the issue is rooted deeper than just a mismatched product fix. It means either:

- You're getting in their way.
- You're creating an opportunity for them to get in their own way.

Customers already know they want to take action—that's why they're here in the first place. The reason for that is rooted in a psychological principle called action bias, which states that when given the option, people would rather take action instead of doing nothing.[59]

Said another way, people want to be able to convert, and it is your responsibility to make that as easy as possible by removing roadblocks along the way. At the end of the day, no one wants to go home empty-handed.

---

59   https://www.ncbi.nlm.nih.gov/pmc/articles/PMC8515773

And just like with any other heuristic shortcut, this can work in a customer's favor or cause them to make a poor decision. Action bias can sometimes lead customers down a path of impulsive and ill-considered purchases. But when harnessed correctly, it can also be a powerful tool to guide consumers toward products that genuinely meet their needs.

To do this, you can:

- Get your customers ready to convert with an emotional appeal
- Make them feel like they already own it
- Hit them with a meaningful and well-timed call to action

SOLUTION 1
GET CUSTOMERS *READY* TO CONVERT WITH AN EMOTIONAL APPEAL

Just because customers want to take action doesn't mean they're ready to just dump everything into their cart and give you their credit card information the moment they hit your product page.

When you ask a customer to take action before they're ready, you risk them not taking action at all—which goes against the outcome you're *both* looking for.

The good thing is that you've already been setting a path for them since the moment they entered your website. Everything they've seen up until now sets the tone for how they will perceive your company and how they're going to act.

As we know by now, the facts behind your company or product have little to do with it.

When people make decisions, they often trust their gut. They often don't know why they feel a certain way, just that they feel it—and that's good enough. We pay therapists a lot of money for that deeper introspection, and we often don't waste that time on our purchasing decisions.

While we all like to believe that those gut feelings come from a deep, introspective, and instinctive space deep inside our inner and truest selves, the reality is a little less romantic. We constantly make connections and decisions based on the subtle cues around us. This is called the priming effect.[60]

No matter what the experience is, your customer is going to have a journey on your website. Long or short, successful or frustrating, they will go down some sort of path. By understanding the cues they find along the way, you can shift those cues to usher your customer into the next phase of the buying journey.

---

60 Molden, D. C. (2014). Understanding priming effects in social psychology: What is "social priming" and how does it occur? In D. C. Molden (Ed.), *Understanding priming effects in social psychology* (p. 3–13). The Guilford Press.

DECISION-MAKING AND CONVERSIONS

Consider the bakery that leaves its doors open, allowing the aroma of freshly baked goods to entice passersby or a retail store that plays upbeat, energetic music to lift their customers' mood while also subtly encouraging them to move more quickly through the aisles and make their purchase.

You can apply the same concept to a website. Its imagery, design, and language prime customers to make certain assumptions or take specific actions. There is usually a single turning point that brings customers from the Decision-Making Phase to the Conversion Phase, and you have to intentionally shape that pivot. If you leave it solely to the customer, you may miss out.

Priming doesn't create emotions or spur action out of thin air—it pulls upon the experiences your customers already have by using the words and images they're familiar with to associate your product with an emotion or action you'd like them to take.

Addressing the emotional impact of a website is crucial, especially because it's not a direct one-on-one interaction with a customer. If there's one key takeaway behind the psychological principles listed in this book, it's this: what *feels* true is often more real than what is actually true. This is called the affect heuristic, which is the idea that people put more stock into emotional appeals than logic.[61]

---

61 Finucane, M. L., Alhakami, A., Slovic, P., & Johnson, S. M. (2000)....

This is especially true in environments where human interaction is minimal or nonexistent (like the internet). Along a customer's digital journey, this means that creating an emotional connection with the visitor can be a pivotal factor in their decision-making.

When customers engage with a website, their emotional response can dictate their perception of the company, often more than factual information. This is because emotions are immediate and powerful, shaping our reality in a way that facts alone may not.

A website that successfully evokes a sense of trust and security is more likely to convert visitors into customers than one that presents a cold aggregation of facts and figures.

This shift has nothing to do with manipulating customers to feel a certain way—it's about showcasing your products and services in a way that helps customers connect the product to the feelings it'll produce when they have it.

Imagine landing on the homepage and being greeted by a high-definition image of a steaming mug of coffee with morning sunlight casting a warm glow over a rustic table. The mug sits beside a bag of coffee beans, which are spilling out onto the table. You can almost smell the rich aroma. This imagery

---

...The affect heuristic in judgments of risks and benefits. *Journal of Behavioral Decision Making*, 13(1), 1–17. doi: 10.1002/(SICI)1099-0771(200001/03)13

taps into the sensory experiences associated with enjoying a fresh brew, evoking a feeling of warmth, comfort, and the invigorating start to a day.

That's how you get customers to connect.

In the realm of SaaS, this approach of creating a sensory-rich experience takes on an even more significant role. Instead of tangible products, SaaS offers digital solutions, but the principle remains the same: show customers the transformative impact the service will have on their lives or businesses. Picture a user interface that's not just functional but also visually engaging, with interactive elements that simulate the ease and efficiency the software brings.

For instance, a project management tool could showcase a dashboard that organizes tasks seamlessly, visually representing the peace of mind and productivity it promises. Accompanying this are images of teams engaged in collaborative work, highlighting real-time communication and coordination. These visuals might include team members brainstorming in a vibrant, modern workspace or presenting polished, completed projects in a boardroom, emphasizing the tool's role in fostering teamwork and achieving successful outcomes.

Through imagery, potential customers can almost feel the relief of streamlined workflows and the satisfaction of enhanced productivity. This is the power of connecting customers with the experiential value of a service, making it tangible in their minds before they even subscribe.

The idea that sales is an emotional process isn't new, but a lazy approach turns emotional appeal into company-focused fluff.

Let's take founder stories, for example. So many companies lead off with their founders' stories early in the decision-making process, as if their CEO's personal journey will be the deciding factor in whether customers buy the product.

Simply sharing the founder's vision and struggles to succeed doesn't move the needle. Every company started with a vision, and businesses rarely succeed without some struggle along the way. That company-focused message comes off as an ego boost for the founder more than anything. It doesn't do anything to advance the customer's goal.

Don't tell customers how the company started. Show them that the founder's origins put them in a unique position to understand their goals and needs right now.

Here's a scenario: the founder is a marathon runner who could never find the perfect running leggings—they were too tight, they chafed, or they simply weren't supportive enough. The store's homepage could feature a powerful image of the founder crossing a marathon finish line, a look of both triumph and relief on their face, wearing the flagship product: the Ultimate Comfort Leggings.

DECISION-MAKING AND CONVERSIONS

Accompanying this image could be a brief caption that tells a story: "Our founder created these leggings after running five marathons in subpar gear. Your comfort is our mission because we've been there."

Focus on the impact, not the action.

This makes it even more powerful to highlight the common pain points your company solves throughout the digital journey. This is where the customer discovers you not only recognize their issues, but you also have the perfect fix ready to go.

In e-commerce especially, nostalgia is a perfect example of evoking a positive emotion around a product. These memories often include times spent with friends or family, which taps into a sense of social connectedness. If you harness it correctly, this connectedness can unconsciously extend to your company.

This can be done in subtle ways though copy, imagery, or even product bundles. Let's imagine your company sells barbecues and other outdoor cooking gadgets. Customers typically intend for this to be a long-term purchase, so they need straightforward photos and in-depth product information to make a decision.

But of course, this purchase goes deeper than that.

They don't just want an appliance that sits on their patio year-round—they want the experience of having a barbeque at their house. That means imagery of families gathered outside on warm summer nights with plates of grilled chicken and corn on the cob covered in tinfoil and a fire pit in the background.

But keep in mind, while it's critical to consider the emotional aspects of customer interaction, it's also important to know where to draw the line. Avoid strategies that could negatively affect customers emotionally, as this not only compromises ethical standards but could also have long-term negative impacts on the company.

The worst offender: an overreliance on a customer's fear of missing out.

There can be a benefit to creating a sense of urgency on your website to encourage conversions. Tactics such as displaying "last one" signs or implementing countdown timers for limited-time offers can be effective in driving consumer decisions. However, honesty is paramount in such cases. Your site visitor may see a "last one" banner on a product, but they aren't ready yet and decide to take their chances. When they come back a week later and see the same banner, they're going to start wondering if it's really true.

Unfortunately, many companies are still of the same old mindset. Except now, customers are smarter. Some websites employ deceptive practices, such as resetting countdown timers upon page refresh or falsely extending sales. These tactics can erode trust and have a negative impact on the customer experience. Think about the last time you got an email that

DECISION-MAKING AND CONVERSIONS

said, "Surprise! Sale Extended!" You knew it wasn't. It was always scheduled to end on this "new" date, and many customers will see right through it.

Sure, this may still work at times, but why invoke negative emotions to make the sale when there are so many opportunities to guide customers through positive experiences? Conversions at the price of your customers' trust aren't worth it in the long run.

Let's See This in Action

When the urgency is genuine, it can be done right. At The Good, we worked with a high-end backyard and camping products company. They found their products frequently out of stock, and we saw an opportunity rather than a problem.

To keep their customers engaged, we implemented a "notify me" feature, ensuring they'd be the first to know when their favorite items were available again. We also made sure the product descriptions highlighted the premium quality of their gear, explaining that this craftsmanship was why they might have to wait sometimes.

We also considered showing the number of items left to help with transparency during shipping delays.

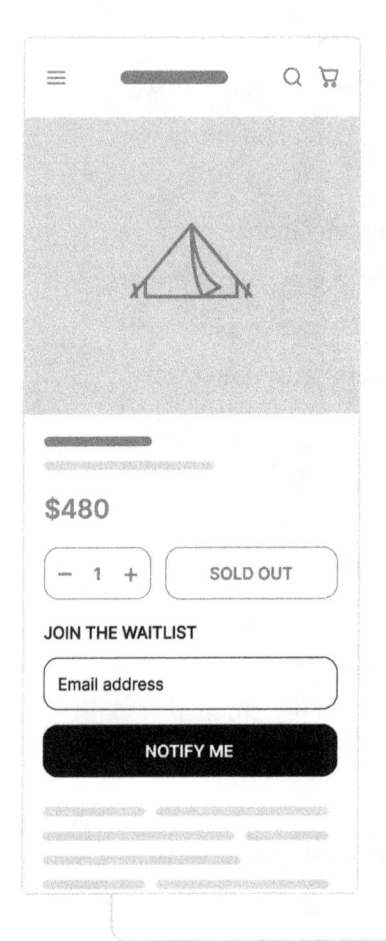

This way, when they alerted their customers that an item was back in stock, those customers were ready to act fast, often clearing out the stock in days.

This approach underscored the difference between creating false urgency and the authentic urgency that comes with high-demand, quality products.

Customers don't like the uncertainty of not being able to acquire a needed product. This ensures that when genuine scarcity arises, they are able to recognize the urgency and make timely decisions.

SOLUTION 2
HIT THEM WITH A MEANINGFUL AND WELL-TIMED CALL TO ACTION

Once your customer feels confident in their decision, it's time to turn that emotion into action.

DECISION-MAKING AND CONVERSIONS

Some tactics include using specific wording to do just that, such as the bye-now effect. It's so straightforward that it doesn't seem like it should work, but it does. Research suggests that merely reading the word "bye" nudges us to think of "buy."[62]

Think of it as the first subtle shift toward your call to action. The best place to test this tactic is in product descriptions, where the customer already has a sufficient understanding of the product at hand, and now they just need a gentle push toward taking action. The customer probably won't even realize why they suddenly feel more inclined to make a purchase—they might chalk it up to "gut feeling."

That's when you hit them with a powerful call to action. The call-to-action button (or other CTA link) is a prompt to take the next step toward becoming a new customer or repeat buyer.[63]

So many companies assume that simply telling customers to convert is enough to mark "CTA" off their digital checklist. The best calls to action, though, don't tell customers what to do—they empower customers to take action on a decision they've made for themselves.

It's a very subtle difference, and one that will play out differently for every single company. No single CTA formula

---

62 Davis, D. F., & Herr, P. M. (2014). From bye to buy: Homophones as a phonological route to priming. *Journal of Consumer Research*, 40(6), 1063-1077. https://doi.org/10.1086/673960
63 https://thegood.com/insights/call-to-action-tactics

works best for everyone or in every situation, but there is one foundational process that will simplify the process dramatically: test, observe the results, and then test again.

When you do, consider these best practices:

- **Use context to set the stage and prepare your visitor to click.** The most effective CTAs are aided by strong context. Make that next step so obvious your customers are eager to take it.

- **Make the CTA stand out plainly so visitors aren't confused about what to do next.** For best results, use just one CTA per page. Clutter and multiple calls to action can send visitors into cognitive overload. When they aren't sure what to do next, chances are high they won't do anything.

- **Be direct.** Don't try to hide the fact that you want them to buy something. When you are sure the products or services you're selling will provide genuine benefits to the right prospects, your visitors will be more confident about buying from you.

- **Keep it personal to build trust and affinity.** Shoppers don't want to feel manipulated or bamboozled. Speak to them in plain language. Use copy on your CTA buttons that clearly conveys the benefit of taking the action and tells visitors what to expect next.

- **Give shoppers a reason to act *now*.** If supplies are limited (and they always are), don't hesitate to say so. If the price will soon be increasing, tell visitors. Each visit presents an opportunity.

- **Experiment with your CTA buttons or links to find out what works best for your particular audience.** Use copy that reflects the benefit of responding to the CTA. CTAs like "Find Out More" or "Show Me How" may be appropriate to start. Near the end of the journey, "Move to Cart" and "Buy Now" could be the best choices.

SOLUTION 3
MAKE CUSTOMERS FEEL LIKE THEY ALREADY OWN IT THROUGH CUSTOMIZATION

Once that emotional connection is in place, you can take it a step further by not just making them feel ready to purchase, but by making them feel as if they already own it.

The more opportunity customers have to create the product or experience for themselves, the stronger sense of ownership they'll have over it when it's in their possession. This is due to a psychological principle called the Ikea effect, which shows that people feel more attached to items they've created themselves.[64]

This effect is aptly named—any time you bring something home from Ikea, you know you're about to spend the next

---

64 Norton, M. I., Mochon, D., & Ariely, D. (2012). The IKEA effect: When labor leads to love. *Journal of Consumer Psychology*, 22(3), 453-460. https://doi.org/10.1016/j.jcps.2011.08.002

thirty minutes to four hours assembling it. But once that last screw is in place, you feel a deeper satisfaction with that purchase because you made it with your own two hands. That same feeling can be translated to the online digital experience through personalization.

If your company offers a highly customizable product, this is an effect you're likely already tapping into without knowing it. Take, for example, a personalized planner and a company that allows users to create a yearly planner that is 100 percent customized to their specific needs. Customers can start with a premade template and customize from there, or they can build it from scratch.

Either way, the intentional effort and time put into assembling the product gives customers a sense of ownership before they even order it. After all, now that you know you have an item that is exactly what you need, why would you bother to search anywhere else for an option that may cover some of your needs but not all?

This applies to e-commerce and SaaS companies in the same way—the feeling of ownership after creating a 100 percent personalized planner would be just as strong if it were a digital planner instead.

The benefit of customization is that it further strengthens the user experience by inviting customers to become partners in the product creation process. The better the user experience, the better the opportunity for conversions and building customer loyalty.[65]

Not only does it increase the chance of conversions, but it also can increase the value of the conversion itself. In fact, research from Deloitte reveals that one in five customers who want personalized products are willing to pay a 20 percent premium for them, and more than half of consumers want personalization and customization options.[66] If your company can offer product customization at scale, your customers will buy more often and possibly even spend more.

While a good first step, offering customized products online isn't enough on its own to increase conversions in most cases.

Here's how you can take it one step further and create a 360-degree personalized experience:

- **Use data to personalize more than just products.** Digital also connects the experience a customer has in-store or on-site to the continued company relationship by providing more interesting content and follow-up offers based on past purchasing behavior.

---

65 https://thegood.com/insights/product-customization/
66 chrome-extension://efaidnbmnnnibpcajpcglclefindmkaj/https://www2.deloitte.com/content/dam/Deloitte/ch/Documents/consumer-business/ch-en-consumer-business-made-to-order-consumer-review.pdf

Offering them meaningful personalization with customization of products, targeted offers, and loyalty rewards gives customers a better experience.

- **Anticipate future needs.** You can use data on what customers are choosing for customization options to anticipate their needs and preferences in the future.

  For example, let's say you notice that pastel shades are suddenly a super-popular choice in custom orders for shoes, particularly pink and blue together. You can use this information to create a new non-custom product in pastel pink and blue and see if they take off as a new trend.

  Or maybe your customer service team noticed that many current users reach out requesting more customization options within the service. It may be time to turn a one-size-fits-all software into a more adaptable, use-tailored solution that allows for individual customization.

- **Test and improve the customer experience.** Increased ease and functionality means you can scale your customization options, even for smaller items. Your buyers get a better customer experience without requiring as much in-person customer service assistance, which is a win all around.

It's important to note that transparency around pricing and shipping times is important, too, as customization typically increases both. Make sure you include these details before, during, and after checkout. We have found that using benefits-focused language like "made for you" can help ease the pain of a longer wait time.

QUESTION:
WHAT SHOULD I CHOOSE?

Up until this point, the online shopping experience involves a heavy cognitive load with customers making small, unconscious decisions throughout the entire journey. But now, that decision is very much a conscious one.

Once a customer has zeroed in on a particular company that resonates with them and appears to meet their needs, now comes the primary decision: "What should I choose?" This question becomes particularly important when a company offers multiple versions of the same product.

Imagine you've spent hours researching and have landed on a company that aligns perfectly with your requirements, whether it's in terms of quality, sustainability, or style. You're almost at the finish line, but now you're faced with a lineup of similar products with varying features. Each version offers something slightly different—be it color options, size variations, or even nuanced functional differences like battery life or fabric material.

Instead of putting that choice on the customer, you have the opportunity to guide them toward the right decision. Because while customers will evaluate information and logic to make that decision, they often don't understand that information and logic is way more subjective than they thought—which can even lead them to make the wrong decision (or no decision at all).

To combat this, you need to:

- Remove barriers
- Compare wisely
- Offer simplified solutions

SOLUTION 1:
REMOVE ALL BARRIERS THAT MAKE THE CHOICE MORE COMPLICATED

Before you help your customer make the right decision, you need to make sure they feel comfortable entering that process in the first place. This is where choice overload comes back into play. When faced with too many options and no clear path to choose, customers often feel overwhelmed by their options. The moment that overwhelm hits, they're much more likely to leave the website completely—and even more likely not to make it back.

Offering an abundance of options may seem like a value-add for attracting customers, but in the end, you're ultimately lowering the value of every option across the board. The more choices customers have, the harder it is to make a decision, and the less confident they are in their decision. Then, because there are so many options available, people expect that that product they choose will be the absolute perfect fit. This ultimately leads to disappointment that has little to do with company, product, or experience—and everything to do with their perception of it.

Consumers want a balance—they want enough options to feel like they're making a choice but not so many that they become paralyzed by the decision-making process. Otherwise, they're thrown into a feeling called decision fatigue.[67] The more decisions we make in a day, the more mentally drained we feel, and the worse decisions we make.

Look at Steve Jobs and his iconic black turtleneck. He wore the same outfit every day to eliminate one more decision from his day, thereby saving mental energy for more impactful choices.

This principle applies to the digital journey as well. Most people are already busy with work and life in general. The

---

67 Lamothe, C. (2019, October 03). Decision Fatigue: What It Is and How to Avoid It. Retrieved from https://www.healthline.com/health/decision-fatigue

DECISION-MAKING AND CONVERSIONS

last thing they want when they visit your website is to be confronted with ambiguity or a barrage of questions. If they can make, for example, just three decisions on your website and then proceed to checkout, they're going to have a far more satisfying experience compared to other companies that ask them to jump through hoops or sift through tons of information.

When you evaluate how to reduce decision fatigue for your customers, think of it as the difference between navigating a maze and going on a hike. In a maze, you find yourself stopped in your tracks every few turns. You have to carefully weigh which way to turn on your way toward the end goal (the exit), even though you only have a vague idea where that exit is. On a hike, you follow a predetermined path through the experience. Sure, you can hop off the path if you want to, but you should be able to easily find your way back. When paths cross or detour and you have to make a choice, there's a clear map that shows exactly where the path will take you.

By reducing the number of decisions a consumer needs to make on your website before reaching the purchase stage, you create a guided process that increases your chances of closing the deal. The fewer choices in their pathway, the better.

To do that, you need to incorporate your company's version of a trail map along the journey. You can do this through a number of different elements.

Curated Collections

Displaying handpicked items in a "Recommended For You" section can guide consumers toward making a choice without sifting through endless options.

Let's See This in Action

Your customers want unlimited options—but only a few choices.

	Category	**Category**	Category	Category	
Subcategory	**Subcategory**	**Subcategory**	**Subcategory**		
Subcategory	Subcategory	Subcategory	Subcategory		
Subcategory	Subcategory	Subcategory	Subcategory		
Subcategory	Subcategory	View All Subcategory	Subcategory		
View All Subcategory	Subcategory		View All Subcategory		
	View All Subcategory				

For one of our clients at The Good, we saw that showing all product categories under one big "Shop" category in the main navigation seemed like it would help users find what they were looking for. But in practice, the overwhelming options actually made search more difficult.

So instead, we displayed the categories themselves in the main navigation, which let users quickly choose what they were looking for.

DECISION-MAKING AND CONVERSIONS

> We started them with just four choices so that they wouldn't get lost before they even started. From there, they could take control of their own search. They could scan more specific offerings to find the perfect match.
>
> The options are now more visible, especially for new users who don't yet know their way around the website—resulting in a 4.25 percent conversion rate increase.

Product Comparison Charts

Implement a feature that allows customers to compare similar products side by side, clearly highlighting differences in features, price, and customer reviews.

> **Let's See This in Action**
>
> When you offer multiple options for a similar product, customers can't make a decision without all the information in front of them.
>
> This business offers four courses. User movement maps showed high engagement on the overall course chart, which compares options side by side.
>
> Engagement is good. Too much engagement without moving on to a conversion can be a sign that

something is wrong. The content may have been different, but visually, it was all the same—forcing people to study the chart carefully to fully understand it.

So how did our team at The Good fix it? By adding in visual cues as well as social proof in the form of the "Most Popular" banner, we made the content more visually interesting and easier to differentiate.

Customers felt guided, which led to a 36 percent conversion rate increase.

Quality Callouts

If you know that your customer is looking for the "Perfect Gift" or "Most Popular" option, it's okay to just go ahead and tell

DECISION-MAKING AND CONVERSIONS

them that. Or better yet, they might be looking for exactly that but not know it right away.

Quick View Option

A "Quick View" feature allows shoppers to see essential product information without navigating away from the main page, minimizing the steps needed to make a decision.

Search Filters

Well-designed filters for categories, sizes, colors, and price ranges can considerably reduce the number of choices a customer has to wade through.

Clear and Easy CTAs

Use clear and concise call-to-action buttons such as "buy now" or "add to cart" that guide users toward making a definitive choice.

Let's See This in Action

For this client, we came into this test with one goal: to streamline conversions. How could users go straight from the product page to checkout? By adding a "Quick Add" button.

By cutting out an extra step (clicking to go to an individual product page before adding to cart), the barrier to buy was that much smaller, making them more likely to convert.

This one change was suddenly worth $257,000 in revenue.

But here's your question: will it work for me? The answer? Maybe!

As always, it depends. Quick add works for this particular client because:

1. Many sales come from returning customers who already know what they're looking to buy when they come onto the website
2. Their products don't require much more additional selling on their own
3. There's little need to explore individual details or educate customers further

But if you have a product that's incredibly specialized, with a lot of details customers need to know in order to

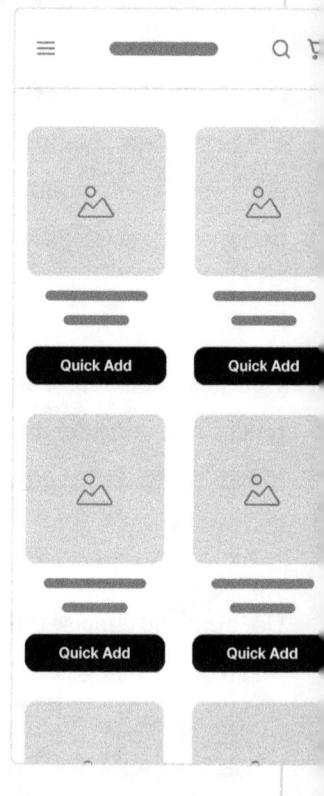

DECISION-MAKING AND CONVERSIONS

make a purchase decision, a "quick add" button may be counterproductive. You may be asking users to make a purchase decision before they're ready.

This is the type of test that may increase conversions dramatically for one website while totally tanking on another.

Simplified Product Descriptions

Overly complex or detailed product descriptions can leave consumers feeling overwhelmed, making it harder for them to make a final choice. Focus on the most important pieces of information a customer needs to hit "add to cart."

Let's See This in Action

Narrowing focus without leaving out key information is a tricky balance. For this client, different variations of the same service were being presented almost as completely different options.

The customer had to work harder to see what was the same and what was different—which made it more difficult to choose.

First, we focused on the key elements of the service because that was what was going to make the sale. Then, by removing less relevant differences, we were left with only the differences in date.

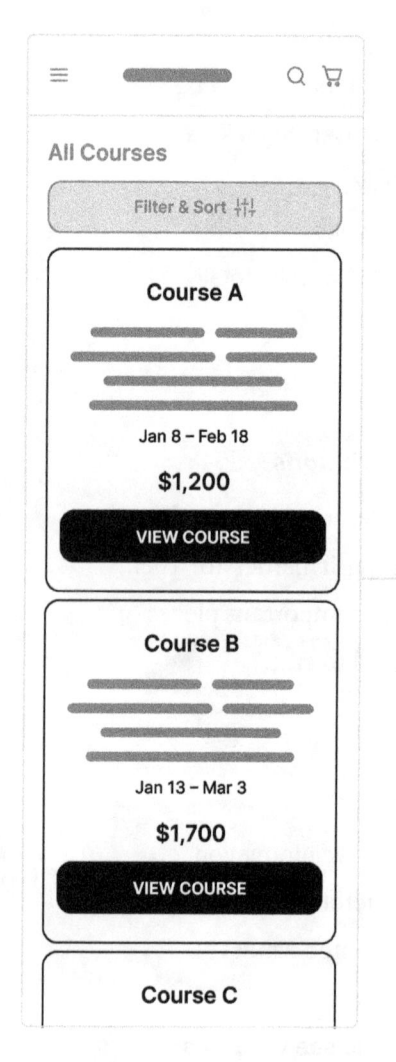

Once they were sold on the service, customers only had one choice to make—when to start.

The results? A 142 percent conversion rate increase.

In-Cart Suggestions

While it's tempting to encourage add-ons, these should be kept minimal and relevant. Bombarding customers with too many additional options can cause decision overload.

Save for Later

Allow customers the option to save items for future consideration. This not only encourages return visits but also gives them the space to decide without feeling overwhelmed.

Checkout Simplification

The fewer clicks and pages between the cart and final purchase, the less likely the customer will abandon the cart due to decision fatigue.

By tracking metrics like high bounce rates and cart abandonment, companies can gauge the effectiveness of these strategies in mitigating decision fatigue. Remember, the goal is not just to sell but to sell the right thing, aligning with the customer's actual needs and preventing post-purchase regrets.

Putting It All Together

The easier it is for a customer to make a decision and act on it, the more likely they are to do just that. The key is to reduce choices along the path, opting to guide them instead. You can do this through:

- Curated collections
- Product comparison charts
- Quality callouts
- Quick view option
- Search filters
- Clear and easy CTAs
- Simplified product descriptions
- In-cart suggestions

- Save for later
- Checkout simplication

SOLUTION 2:
COMPARE WISELY

When making a final decision, most customers are going to compare products side by side to determine the best fit for them. But because of heuristics, the way they approach that comparison can ultimately impact their decision.

It's possible that you might inadvertently create a problem in this scenario. The potential issue arises when customers find themselves in a position of comparing closely related products or services—as anyone who has ever tried to buy a new microwave would know.

Maybe one microwave has 800 watts power but another has far more heating options. Being able to compare options side by side gives a visual view of the key deciding factors that build trust. That's where comparison charts come into play to list out the features and benefits in a clear, streamlined way.[68]

The main elements in a product comparison are:

---

68 https://thegood.com/insights/product-comparison/

DECISION-MAKING AND CONVERSIONS

- **Product description and details:** What are the specifications, and how do they differ from other similar products? For example, one microwave might be cordless and smaller than another.

- **Product features:** What add-ons does the product have that others don't? For example, one microwave might have dozens of preset options, while another only has a button for popcorn.

- **Product benefits:** How will the product make life easier for the consumer, and how does this differ from other similar products? For example, one microwave may have removable parts for easy cleaning.

Product comparison slots in at the consideration stage of the sales cycle, helping shoppers identify the best-fit product for them in a quick glance and, therefore, boosting conversions.

How you present your product comparison page will depend on the type of product or service you're selling and the kind of people you're selling to. However, there are some simple strategies to keep in mind so you match customer expectations:

- **Include images:** Display images of the products you're comparing so customers know exactly which items you're talking about.

- **Make it visual:** As well as product or software photos, include illustrations and other visual elements to maintain reader attention.

- **Stick to less than five products:** Don't overwhelm customers with tons of options. Instead, stick to two to five for best results.

- **Consider shopper interests:** Bring the features that are most important to your customers to the top of the chart.

- **Keep it simple:** Keep text to a minimum, and avoid listing complicated features that will confuse customers.

- **Include social proof:** Add ratings and customer reviews to your chart so shoppers can see how previous buyers felt about the product.

Comparisons are often a crucial part of the decision-making process. The better you can facilitate that comparison task

DECISION-MAKING AND CONVERSIONS

for your customers, the easier it will be for them to make a decision. Instead of opening hundreds of tabs and flicking between them all, customers will have the information they need all in one place.

But just giving the customer the information they need isn't enough—you can frame it in a way that clearly leads them though the customer journey path so they don't get stuck at a juncture making decisions between elements they can barely differentiate. The way your customer approaches those comparisons can drastically alter their decision—even if they're looking at the same information.

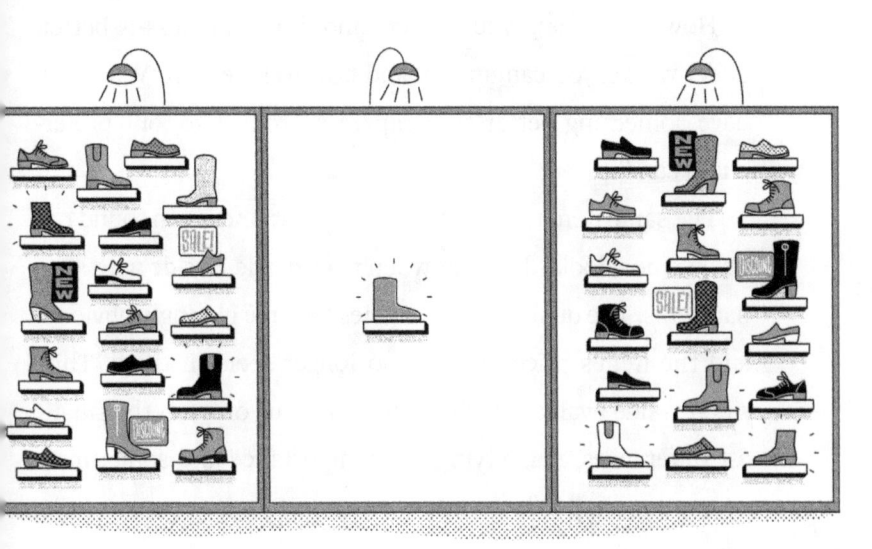

Take, for example, the less-is-better effect. This effect highlights a behavioral quirk where shoppers may actually favor an item of lower quality when it's the only one they're considering, as opposed to comparing it alongside better options.[69]

In other words, we're often drawn to something a bit less impressive when it's the only thing in front of us.

When evaluating products or services one at a time, customers can determine if each meets their needs and let that distinction be enough. It can even increase the perceived value of the offering. Customers may find themselves willing to pay a little extra for an item that, in isolation, appears to be of reasonable value. Without a direct comparison, it's easy to accept the cost as the going rate for such a product.

However, when you understand how the less-is-better effect works, you can understand how to reverse it. When we have something better to compare a product to, our preferences can flip.

The same item that seemed like a great find when found on its own now looks different when placed side by side with similar items. The quality discrepancies become glaringly obvious, and the item's price tag may no longer seem justified. This side-by-side evaluation sharpens our focus on even the smallest differences, magnifying their importance and sometimes prompting snap judgments.

---

69 Tversky, A., & Kahneman, D. (1981). The framing of decisions and the psychology of choice. *Science*, 211(4481), 453-458

DECISION-MAKING AND CONVERSIONS

These decisions might feel right in the moment, but upon reflection, they may not always seem like the best choice.

Does that mean you have a bad product or faulty service on your hands? No, not at all. Actually, it doesn't even mean that the lower-quality option is the worse choice. It could be the absolute perfect fit for your customer, but compared to something with more bells and whistles, all of a sudden it doesn't seem as good.

Let's look at it like you're buying a car. The base model comes with all the standard safety features. But the next tier up comes with surround sound, voice-activated commands, park assist, and a drowsiness detector—all of which are gadgets that you didn't need (or even know existed) when you got to the dealership. If you had just looked at the base model and nothing else, you would have found that it was perfect for what you needed. But now because you compared the two side by side, you know what you're missing—even though you don't need it. Suddenly, that base model doesn't feel worth it compared to the next step up.

To use this principle correctly, you have to know your products and your audience backward and forward. If you have a product that you know perfectly fits your customer needs, let it stand alone. But if you're looking to upsell, particularly when it comes to SaaS products, a comparison may serve you better. The most powerful comparison charts highlight features that are only included in higher-tier options that you may not need but suddenly are willing to pay more for now that you know it's an option.

For example, let's look at PandaDoc, a document signing competitor to DocuSign.[70]

Fair pricing to keep your business growing

Essentials — $19 / month

Business (MOST POPULAR) — $50 / month

Enterprise — Let's talk

It's likely that the majority of customers don't need anything beyond the Essentials package. Had they seen that

---

70   https://thegood.com/insights/saas-pricing-page

option on its own, it would have felt like this package included everything they needed. But now looking at the Business option, suddenly customers may think about how much easier it is to gather the information they need using forms instead of a long email chain or how much more convenient it would be to bulk send documents.

The layout is simple and clean, quickly guiding the user down the page. PandaDoc goes above and beyond, not only by offering extremely transparent pricing, features, and use cases for each plan but by adding a section comparing each plan's more advanced features so customers know everything that's included in any plan.

The key is to incorporate elements that help customers make the decision, such as:

- Clear and detailed information about the features and benefits of each pricing plan
- Toggler for annual plan and monthly plan so users can choose between the two
- Pricing plans with different levels of features and pricing, providing customers with options

Then, they follow up with elements that validate that choice, such as:

- Customer testimonials
- Option to try PandaDoc for free to encourage potential customers to test the product and see its value before purchasing

But keep in mind that while many customers will take all comparison factors into consideration, many won't. This is known as the take-the-best heuristic. Essentially, instead of carefully considering all elements in the decision, customers may make their decision on one key factor.[71]

---

71 *Take-the-best heuristic.* (2020). Psychology Concepts. Retrieved November 12, 2021, from https://www.psychologyconcepts.com/take-the-best-heuristic

DECISION-MAKING AND CONVERSIONS

This mental shortcut lets us zoom in on what really matters to us when picking between products so we can make choices confidently and quickly. Instead of diving into every detail, shoppers tend to pick a standout feature—like price, for example—and let that be the deciding factor.

Getting to the heart of what your customers really value—whether it's cost effectiveness, quality, or the convenience of use—is crucial. This is where you ask yourself, "What is the most important factor for my customers?" Is it price? Quality? Versatility? Ease of use? If you aren't sure, turn to customer reviews to see what is mentioned the most.

Once you narrow it down, highlight that particular quality on your product page to let your customer know that you understand their needs and are here to fill them.

Ideally, the aim should be to prevent customers from reaching that stage where they're faced with the task of comparing two products that are almost indistinguishable. Instead, a well-designed filtering system should be in place. This way, the consumer is seamlessly guided through their journey, and they are directed toward products that are a perfect fit for their needs right from the start.

SOLUTION 3:
OFFER SIMPLIFIED OPTIONS

One of the best, most sales-focused approaches to combating decision fatigue is through packaged options. Enter: the decoy effect.

Businesses often strategically introduce a decoy—a less attractive option—to shift customers' preferences toward a target product they want to promote. The presence of the decoy simplifies the decision-making process by providing an easy-to-reject option, making the target seem more appealing and valuable in comparison.[72]

E-commerce and SaaS industries often employ a decoy in their pricing structures. It involves presenting customers with three variations of a product or service, each with different features and price points. Providing a trio of choices subtly nudges consumers toward the middle-tier option, which typically strikes the best balance between cost and value, aligning with what most shoppers are seeking.

For instance, imagine you're offering a proposal for client services. The first option costs $100,000. It covers roughly 95 percent of the client's needs. The next option might be

---

72 Hendricks, K. (2018, December 7). *The decoy effect: Why you make irrational choices every day (without even knowing it)*. Kent Hendricks. https://kenthendricks.com/decoy-effect

DECISION-MAKING AND CONVERSIONS

$120,000. It offers everything the client is looking for, plus a little extra, at just a smidge above their budget. The price isn't much higher than the first, but it offers significantly more value. Alongside it, you introduce a premium option for $1 million. It covers outrageous features that no one will ever need, and they're never going to go for it. But that's okay. The intention isn't necessarily to sell this top-tier option, but rather to use it as a benchmark. This makes the middle-tier option—priced just above the client's expected budget—appear more attractive.

This approach also restores the customer's perceived sense of agency over their buying decisions. They may opt for something a bit beyond what they need, but the key is that it feels like a choice they've made independently.

Think of it like the Goldilocks effect. One package is way too big. Another doesn't cover enough. But the one in the middle? That's just right. Does it mean that the middle package is actually just right? Maybe. Maybe not. But it *feels* right—and that's the difference.

That feeling of fulfilling our needs is something we chase in all aspects of our lives. If you survey coffee shops around the country, for example, and ask the right amount of coffee for the average customer, the answer isn't going to be measured in ounces. The right amount of coffee is whatever is served in a medium-sized cup.

Let's say you have 8 oz, 12 oz, and 16 oz options—it's pretty likely that the majority of customers will choose the 12 oz. Why? Because a smaller size might mean they don't get the full caffeine boost they're looking for, but a bigger size might send them into jitters for the rest of the afternoon.

But let's say we change the sizes, and now you serve 6 oz, 8 oz, and 12 oz options. Suddenly, 8 oz is the right amount of coffee because 6 oz might not be enough, but 12 oz seems like too much. Whereas before, 8 oz was too little and 12 oz was just right. It's not that the medium option *is* right—it's that the medium option *feels* right.

When you understand this key concept, you can incorporate it into your pricing in a way that both helps your customers make an easier decision and increases your revenue. It's a win all around.

In product offerings, especially in the SaaS industry, you often see a tiered comparison chart with checkboxes indicating features available in each package. This visual approach can help, but it can also become overwhelming if not done right. The low option might have twenty out of one hundred checkboxes filled, while the middle option has seventy, and the high-end, expensive option has all the checkboxes ticked.

Then, companies can focus on highlighting the unique benefits of each package through a clear description at the top. Consumers usually want to know what specific problems each package solves. This allows them to easily understand the differences between the options and make a more informed choice.

For example, Mailchimp, an email marketing platform, employs a pricing structure that separates each pricing table by use case. However, Mailchimp doesn't use separate landing pages, but rather uses toggle options so users can select what they want.

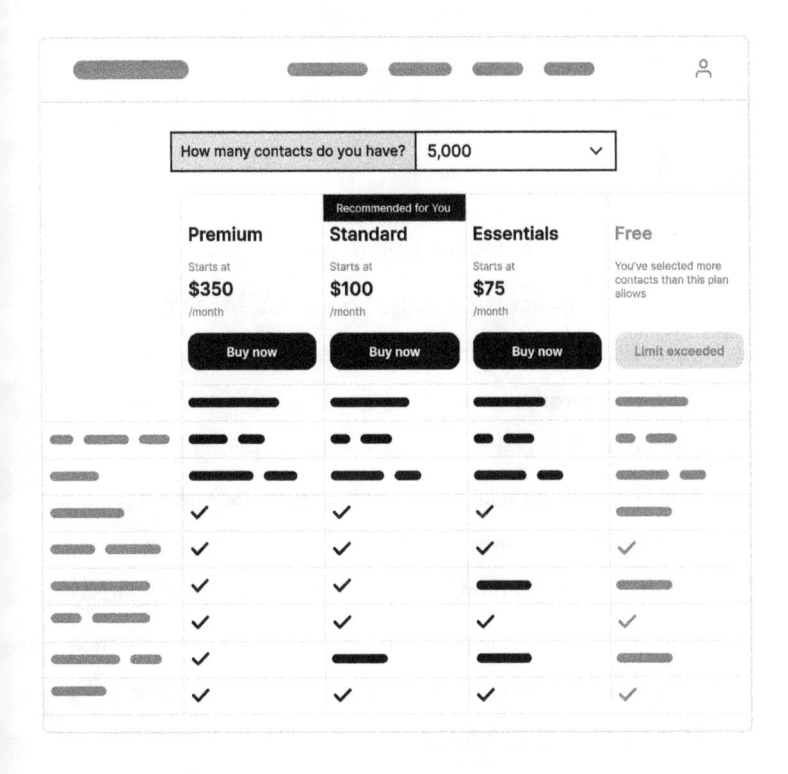

Cleverly, Mailchimp includes a drop-down menu that allows customers to select how many emails they need, which gives them the pricing immediately and puts the customer at ease that there are no hidden fees or complicated pricing structures.

Then, it lists comprehensive features available in each plan, including email sends, users, audiences, customer support, prebuilt templates, and more. So if customers need or want a specific feature, it's easy to find without asking the customer service team.

It's not just about the number of features—it's about the quality and relevance of those features to the consumer's particular needs.

Let's see this in action

A premium boat manufacturer serves as an intriguing case study in tackling both decision fatigue and the Ikea effect from the previous question. Our team at The Good was tasked with revamping their platform, catering to two distinct types of customers. The first type was the affluent buyer seeking a high-quality boat primarily for family activities. They were willing to splurge for the sake of family time, essentially thinking, "If I own a cool boat, my kids will want to spend time with me, and their friends will too."

DECISION-MAKING AND CONVERSIONS

The second type of customer was often the children of these affluent buyers. These kids wanted their dads to invest in a top-of-the-line boat and would even custom design it, down to the stitching color. The platform was built with customization as a key selling point. You could tailor almost every feature of the boat, from the paint to the stitching color.

This presented us with a challenge: how to cater to both the customer who wants full customization and the one who simply wants a quality boat without having to decide on every little detail.

We solved this dilemma by creating package options that were preset with popular features and colors. These package options catered to affluent buyers who didn't have the time or inclination to customize every single feature. Plus, these prepackaged boats were readily available at most dealers, meaning customers could make a same-day purchase as opposed to waiting three to six months for a custom-built model.

On the flip side, we noticed that it was usually the fifteen- to eighteen-year-old kids who engaged most with the customization features. These teenagers would spend time customizing the boat online but often hit a dead end because they weren't the ones with the purchasing power. Previously, their customized designs would be lost when they left the website.

We remedied this by allowing them to save their customizations through a unique URL. They could then forward this URL and a quote from a dealer to their parents. This made it easy for the parents to understand what their kids wanted and make the purchase if they chose to.

Reducing the number of decisions a customer has to make can alleviate the fear of making a mistake. The idea is to offer compromises that people can feel comfortable with. For instance, some buyers may not want to invest time in customizing every aspect of a purchase. They'd prefer something off the shelf, and they're completely okay with that.

Understanding the core reason for a purchase is also crucial. Take the boat-buying example: if the primary goal is to spend quality family time, then the color of the seats becomes less significant. The child might care about seat colors to impress friends, but for the parent, it's all about family time. That's the "job to be done," and as long as that job is accomplished, the other details may not matter as much.

DECISION-MAKING AND CONVERSIONS

QUESTION:
WHAT'S IN IT FOR ME?

In every transaction, the customer expects to get something back in return for their time and money. You'd think that the product or service itself should be reward enough. But when competing against countless other companies offering the same or similar solutions, many feel the need to sweeten the pot.

They do this through discounts, promotions, and rewards.

We previously discussed discounts when used as a first impression. When introduced in the Information-Gathering Phase, a discount acts as an anchor point. When the first thing a customer sees is a discount, it's going to impact every interaction along the rest of their digital journey. This could cause them to see non-discounted items as "not worth it," or worse, regard your company as a "discount brand."

But the same elements or strategies can be employed in different ways, for different reasons throughout your customer journey. Utilizing discounts, promotions, and rewards in the right way in the right moment of the digital journey can be extremely beneficial.

Incentivization right at the brink of conversion is like a spark that can ignite a shopper's decision-making process. By offering rewards such as discounts or special bonuses, companies activate the reward-seeking circuits in our brains, leading us to engage in behavior that could earn us those rewards.[73] It's like you're leaving behind a trail of breadcrumbs that lead your customers right to the checkout page.

The problem is, most companies don't implement incentives the right way. They understand this part of the theory and stop there. They lure customers in with incentives without realizing the true cost of how much they're giving away.

When evaluating psychological tactics, it's important to understand that some tactics can guide your customers forward positively, while others push them forward, but at a cost (and sometimes in the wrong direction). If you have a truly optimized customer, you shouldn't need to tap into the part of your customer's brain that make them want your products or services. They should already want it.

So think of incentives like discounts as a shortcut—one that you do not need if the journey is strong leading to this point, and one that is ultimately going to do more harm than good.

If shoppers come to expect constant sales or discounts, they might start to wonder about the actual value. A company that constantly offers discounts may start to feel less like a

---

73 Azab, M. (2017, February 23). *The neuroscience of wanting and pleasure*. Psychology Today. https://www.psychologytoday.com/ca/blog/neuroscience-in-everyday-life/201702/the-neuroscience-wanting-and-pleasure

DECISION-MAKING AND CONVERSIONS

treat and more like a baseline expectation. Now, all of a sudden, you are a bargain company.

Yes, rewards can reinforce behavior we want to see repeated, but offering rewards for everything can lead to a lack of motivation when there's nothing up for grabs.

We don't just see this with discounts. Incentivization can work the opposite way too—a concept we call negative intent shaming. This refers to the practice of a company attempting to dissuade a consumer from taking a particular action the company doesn't want them to take.

An example of this that's quite common is the use of pop-ups offering discounts to incentivize users but also including prompts like "No, I don't like discounts" or "No, I don't want rewards." Instead of allowing users to simply close the window or decline the offer, these pop-ups try to make users feel guilty or hesitant. It's important to note that while this approach might momentarily stop some users and make them reconsider, it doesn't necessarily lead to higher conversion rates.

The core question here is whether you should resort to incentives at all costs, or if you should focus on building a solid foundation for your company and products. In essence, are you trying to incentivize every step of the way, or should you prioritize creating an experience and value that naturally drives conversions?

Incentives, when used excessively, can set unrealistic expectations and erode the value of your company. They might give you short-term gains, but they could also decrease the margin for error and consumer patience.

Consider a company like Apple, which rarely, if ever, offers discounts. While you might find Apple products discounted through resellers, Apple itself maintains its prices. This is because Apple has built a company that offers premium experiences and products. Discounts are not needed to entice customers because the value and loyalty associated with the company are sufficient.

If you find yourself running sale after sale just to increase your conversion rate, it's time to stop. Remember, you can cut your prices to one dollar and send your conversion rate through the roof right now—but you're only going to make a dollar. That type of conversion rate boost isn't really what you're after.

If you want to strengthen your customer journey to increase profits, it's time to invite your customer back into the process. Instead of focusing on facilitating transactions with a faceless entity on the other side of a screen, shift it to fostering trust and genuine interactions.

SOLUTION:
GIVE SOMETHING EXTRA INSTEAD OF TAKING SOMETHING AWAY

At its core, incentivization is all about encouragement. It should motivate shoppers by making them feel valued and excited, not coerced. It's a delicate dance between giving a reason to act and not letting that reason overshadow the action itself.

The key here is to offer a promotion, not a discount. While that might seem like the same thing, they couldn't be more different.

Discounting refers to direct dollar discounts or taking a percentage off your price. Discounts encourage the customer to look at price as the main (perhaps even the only) influence on their decision. Think: a "10% off your first order when you sign up here" pop-up.

Promotions are anything other than slashing the price. Promotions offer additional value to the customers—something they wouldn't otherwise receive. Done well, this incentive can be more compelling than saving a few dollars. Think volume- and loyalty-based perks that give "something extra."

At this point in the digital journey, your customer is already close to making a decision. They've gathered the information they need. They've reviewed the price point, the reviews, and the product details. So you have to think about how this tactic helps reach your goal.

The goal: prompt the customer to take action right now.

When you discount, you're now voluntarily giving a price reduction that your customer didn't expect and likely didn't need. If they came into this journey looking to score a deal, that mission would have led them down a completely different digital path. In the customer's eyes, the discount is nice, but it's not going to be the deciding factor.

Whereas promotions (done well) lead with the value of your product. They account for the values that lead your customer to the tipping point of their decision and then add a little bit of extra incentive to push them toward "check out."

There's a big psychological difference for your customers, which means there's a big difference for your company and revenue too.[74]

Remember, the focus should be on building a company that naturally attracts and retains customers based on the inherent value it offers. By understanding the psychological principles at play here, you can use them to your advantage in a way that works for your customer and your conversions.

Here are a few tactics that will get your customers thinking differently about their purchase.

Add an Element of Surprise

The allure of potential rewards often has a remarkable effect on our online shopping behaviors. This interesting dynamic, known as the motivating uncertainty effect, suggests that the excitement associated with rewards of uncertain value often drives us more than the promise of certain rewards.[75] It's not just about winning—it's the thrilling possibility of "what if" that captivates us and may lead us to invest more in the chase—be it our time, effort, or money.

---

74 https://thegood.com/insights/discounting-for-ecommerce
75 Motivating Uncertainty Effect Definition. *Convertize*. https://tactics.convertize.com/definitions/motivating-uncertainty-effect

- **How companies do it wrong:** To take advantage of this, many companies dive straight into the uncertainty with strategies like the "spin to win," which is hardly ever a good idea. Sure, it might be effective at collecting email addresses, but it also comes across as gimmicky and disrupts the user experience. Not only are you offering discounts, but you're also presenting a whole array of discount options. This sets up a scenario where customers might expect a substantial discount but end up with a smaller one. You're potentially upsetting customers right from the start.

The problem is compounded by the fact that many companies have this "spin to win" feature as the first thing that pops up when users visit their website. There's no delay or opt-out option, so it immediately occupies the entire screen, making it challenging for users to navigate or close without submitting personal information they likely aren't ready to give yet.

The only time it seems to work is when you have someone who is already committed to buying from you, and they're simply excited to get any discount—which means you likely discounted purchases for customers who would have converted at full price.

- **How to do it right:** Want to make uncertainty fun? Offer a mystery bag.

  E-commerce retailer Woot offers a mystery bag for purchase, typically on a weekly or monthly basis, and the catch is that you have no idea what's inside. All they disclose is that the minimum value of the contents is $250. But you can buy it for just $25.

  The key to this strategy's success is that it creates anticipation and excitement for customers, which provides substantial value that reinforces company loyalty. It transforms excess inventory challenges into an experience that customers eagerly await, generating buzz and goodwill without cheapening the company.

SaaS companies often effectively utilize this strategy through gamification, the practice of "gamifying" certain tasks or behaviors by designing user experiences that are more interactive and fun. Tasks that might otherwise be boring or mundane are elevated to improve customer engagement—think apps like Duolingo, which turns language lessons into miniature games and uses badges and levels to indicate when users accomplish a learning goal. Additionally, the app encourages users to set their own daily goals of time spent learning.

While uncertainty can be a great motivator, it only works when customers feel secure enough to take the risk. Foundational elements—such as upfront shipping dates, price protection policies, and clear communication about purchase follow-ups—create a bedrock of trust. This trust allows customers to engage with elements of uncertainty with excitement rather than trepidation.

If the focus is too much on the unpredictability, customers may shy away, preferring clarity and certainty. The aim should be to create a digital experience where the excitement of potential rewards enhances the overall value of the company and its offerings.

Give a Little Extra

Incentivize customers by adding to their purchase, not by discounting what they've already decided to buy. This makes them feel like they're getting the best deal possible, even if it's not much of a deal at all. You can do this though tactics like:

- **Free gift with purchase:** Incentivize customers to make an initial purchase by offering them a complimentary gift when they order now, or offer a free add-on feature for subscribing within the promotional period.

- **BOGO:** Clear out inventory with a "buy one, get one" offer. In e-commerce, you can employ "buy one, get one half off" or "buy one, get one free." In SaaS, "buy one subscription, gift one for free" or "buy one year, get the next free" are great options.

- **Free shipping (for e-commerce):** While this has become a bit of an expectation thanks to Amazon, it's still a compelling incentive—especially on bulky items, like furniture and mattresses, where shipping usually costs an arm and a leg. Variations like free expedited shipping (e.g. two-day over five-day) or free shipping above a certain cart value (to protect your margins) are also effective.

CREATE LOYALTY PROGRAMS

As we've discussed in previous chapters, people often don't buy a product just for the product alone. Often, they buy a product because it says something about who they are or who they want to be. It helps them connect to a larger community outside of themselves. Add value by making them a part of your company, not just a one-time customer.

One of the best ways to do this is through loyalty programs. According to a Wirecard survey, 92 percent of consumers are swayed by rewards some or all of the time.[76] So reward loyal customers with a set of appealing perks that encourage repeat purchases. For example, Mack Weldon offers tiered loyalty rewards. Their CEO, Brian Berger, explained to CNBC, "We have a permanent loyalty program. And that gets us out of the cycle of having to think about promotions and re-training customers in a way we wouldn't want them to behave."

---

[76] https://nmgprod.s3.amazonaws.com/media/files/97/e3/97e3466268a5f6a39748b0acf861188d/asset_file.pdf

OFFER BUNDLES, GIFT CARDS AND CASH BACK

For companies that offer digital products, bundles are the perfect way to offer additional value without impacting the sale's bottom line. Digital product packages, user bundles ("get five additional users for free when you sign up today"), and functionality bundles act as a great push to make customers feel like they're getting more out of a product they were already seriously considering.

The e-commerce realm is a little tricker, when each additional product bundle actually cuts into that product's bottom line. Well, sometimes.

We don't want to trick customers into thinking a product or service is better than it is, but there is a way to create a valuable offer to your customers that, even if they don't use it, still makes them feel as satisfied as if they did—while also putting money in your company's account.

The bundling bias is a cognitive bias that affects how we perceive the value of bundled packages and gift cards. Essentially, it refers to our tendency not to fully utilize all the items or experiences included in a bundle. Customers don't get the full value compared to buying items individually—but they often feel like they did. Businesses often capitalize on this bias by creating bundled packages that make the customer feel like they're getting a greater value than what they came in for.[77]

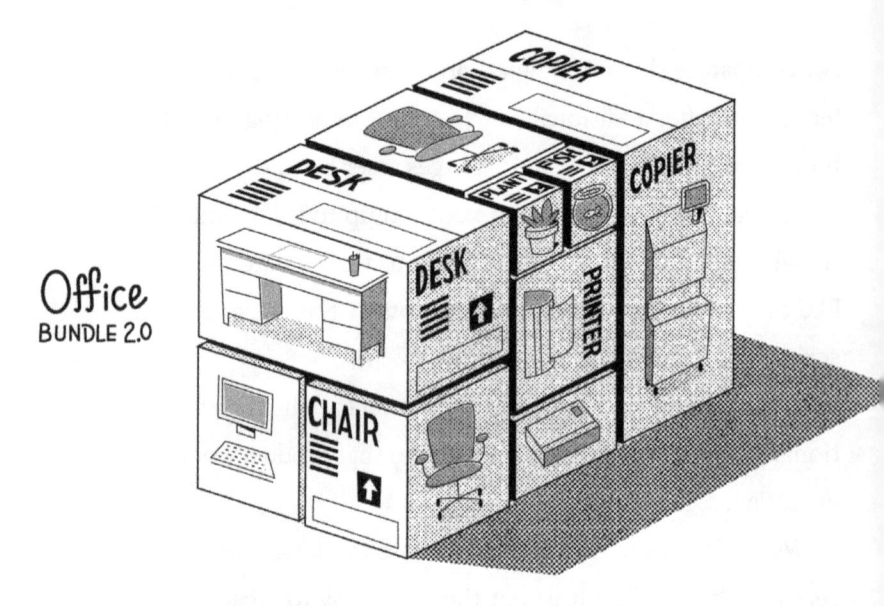

Office
BUNDLE 2.0

There are two ways to look at this: one involves adding value, and the other involves putting the responsibility to use the product to its fullest potential on the customer, not the company.

If you employ this tactic to create a bundle that adds value, you take what you know your customer is looking for and throw in a little bit extra to make them feel like they're getting

---

77   Coglode. (2020, September 22). *The Risk of Bundling*. https://www.coglode.com/gem/the-risk-of-bundling

DECISION-MAKING AND CONVERSIONS

something more—even if they don't need that extra bit. This is especially true for gifts.

Imagine for Father's Day, a wife purchased a detailing package for her husband's car. This package included a number of big containers that he'll use every time he cleans the car. But there are also numerous small containers of products that he might never use. His wife could have just bought the things he needed, and it would have been enough. But the inclusion of these additional, albeit less relevant, items makes it *feel* like a more substantial gift.

In this way, bundling allows you to offer something extra to your customers, enhancing the perceived value of their purchase without resorting to straightforward discounts. It's a strategy that can be highly effective in maintaining your company's integrity while still providing attractive incentives to your customers.

Companies can also take advantage of this through packages. Let's say you join a yoga studio. Classes are $20 each, but a package of ten classes costs $150—that's only $15 dollars per class. Obviously, it's a better deal in the long run, right?

Maybe! This is where customer accountability comes in. Let's say you only make it to seven classes. Because of the bundling bias, the customer is much more likely to rationalize, "Well, I already paid for all the classes, so it doesn't matter if I don't go because the money is already gone." It's a conscious decision on their part, so they still feel like they got the full value, even though now, those seven classes actually cost them $21.43 per class.

Gift cards and cash back work on a similar principle—consumers may not spend the entire value of the card, and retailers benefit from receiving upfront payment. Cash-back programs also play into this bias, with some consumers perceiving cash back as equivalent to discounts.

However, whether these tactics are ethical depends on consumer awareness and intent. If consumers knowingly purchase bundles or gift cards for gifting purposes and are comfortable with not using every component or dollar, it's considered ethical. However, if businesses intentionally deceive consumers with bad bundles or gift cards, it's seen as an unethical practice.

Done right, it adds value for the customer and for your company's bottom line.

Putting It All Together

If you're looking for optimization tactics that convert successfully and sustainably, discounts are not for you. You don't need to give away your company's value to make your customers feel like they're getting more from you. Instead, you can:

- Add an element of surprise
- Give a little extra

- Create loyalty programs
- Offer bundles, gift cards, and cash back

QUESTION
IF I'M WRONG ABOUT THIS CHOICE, WHAT IS THE WORST POSSIBLE OUTCOME I COULD EXPERIENCE?

Purchasing something new can be fun — retail therapy is a real thing, after all. But it can come with a pinch of uncertainty. As customers, we often pause and think, "What if this choice doesn't pan out?" It's a fair question, considering that each purchase is more or less a leap of faith that if you give your personal, private banking information to some faceless entity on the internet, you'll get what you need in return.

This isn't just about buyer's remorse over an ill-fitting sweater or a gadget that doesn't live up to expectations. There's an inherent risk that isn't just about product satisfaction—it's about data security, privacy, and the reliability of the entire transaction.

Many companies understand this hesitancy on the surface. But if you want customers to feel confident in their decisions, you need to know why that hesitation bubbled up in the first place because it goes much deeper than just your run-of-the-mill nerves. There are a few psychological principles at play here that contribute to these feelings.

This lingering doubt in a customer's mind primarily comes from the fear of making the wrong decision. The problem is, the fear of making the wrong decision can actually lead them to—you guessed it—make the wrong decision.

Loss aversion plays a pivotal role in consumer behavior, particularly when it comes to trusting new companies. It's the idea that the discomfort we feel from a loss is more intense than the joy of an equivalent gain.[78] Think of it like this: losing $50 feels more upsetting than the happiness we get from finding $50.

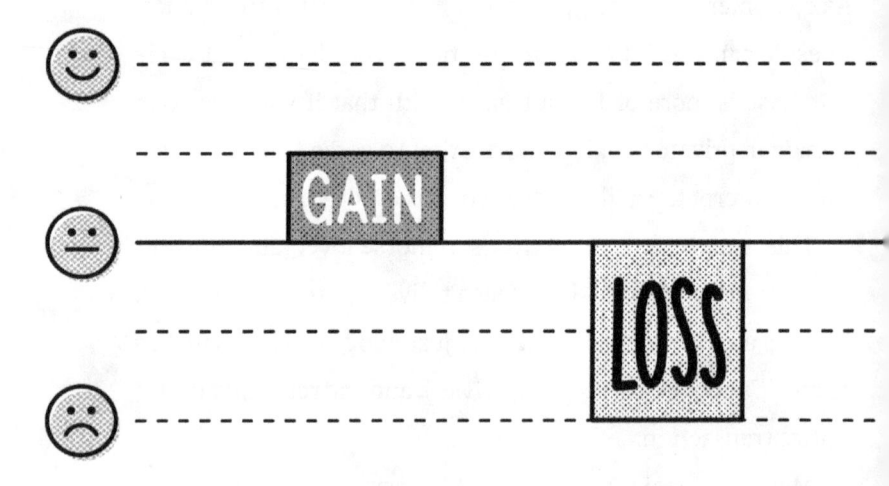

This concept comes into play when customers are considering whether to buy from a new company or try a new

---

78 Kahneman, D., & Tversky, A. (1977). Prospect Theory. An Analysis of Decision Making Under Risk. doi:10.21236/ada045771

DECISION-MAKING AND CONVERSIONS

product. The risk of the unknown can seem daunting. After all, sticking with a familiar product may not be perfect, but it feels safer than trying something that may not work out. It's the classic dilemma of "if it ain't broke, don't fix it." Even if it actually means "this fits the bare minimum requirements, but the other choice might be even worse, so this is fine."

The risk of buying something new that doesn't work (a loss) can be more uncomfortable than the happiness customers get from buying the perfect item.

When you're competing against the big guys, that can feel like a losing battle. It's not, but there's a reason that many customers instinctively lean toward those tried-and-true names, even if they know there are better-quality options out there. Customers often favor what they already know and trust, which can make them hesitant to try new products or companies—this is known as status quo bias.[79] It's a natural preference for the familiar, partly because it's seen as less risky than the unknown.

For those who shop online, this might mean they prefer to stick with well-known retailers and products rather than explore offerings from new companies.

The antidote to this anxiety isn't to downplay or dismiss these legitimate concerns. Rather, it's about acknowledging and addressing them before they even bubble up to the surface of the customer's consciousness. Trust is the cornerstone of conversion.

SOLUTION:
BUILD TRUST AND THEN GUARANTEE IT

The moment doubt casts its shadow on a customer's confidence in a company, it becomes harder to win them over. This is where the power of preemptive reassurance comes into play.

---

79 Samuelson, W. & Zeckhauser, R. (1988). Status quo bias in decision making. *Journal of Risk and Uncertainty*, 1(1), 7–59.

DECISION-MAKING AND CONVERSIONS

Offering guarantees that speak directly to potential worries can be a game-changer.

It all comes back to loss aversion. People naturally avoid losses at all costs, especially when it comes to finances. So when there are risks involved (which, in purchases, there always are), we naturally create shortcuts to evaluate them.

There are a couple of psychological tendencies at play.[80] The first is that people tend to prefer a guaranteed smaller victory over a larger one that's less certain. Let's say a customer is given a choice between a coupon code that offers a modest but guaranteed discount on their current purchase and the opportunity to enter a contest where they could win a much larger discount or even a free product or add-on service. Despite the allure of potentially winning big, most customers will opt for the immediate and certain savings provided by the coupon code. The immediate discount is a sure thing, which appeals to the customer's desire for instant gratification and the avoidance of the disappointment that would come with not winning the contest.

The second is that people are more inclined to risk the possibility of a greater loss rather than face a smaller loss that's more likely to occur. For example, when buying an expensive

---

80 Kahneman, D. (2013). *Thinking, Fast and Slow* (1st Edition). Farrar, Straus and Giroux.

electronic device, they might skip the additional cost of a protection plan that covers any damages or issues, which represents a smaller but more immediate expense. Instead, they gamble with the possibility that if the device breaks or malfunctions after the standard warranty period, they could face a much larger cost for repair or replacement.

Or instead of migrating over to a much more robust software that fulfills all of their needs, they choose to stay with their just-good-enough platform and continue doing other tasks by hand that the new software would otherwise automate for them. Here, there's a significant risk that the customer will spend valuable time and resources migrating data from one platform to another and retraining the team, only to find that the new software doesn't actually perform as well as it should. Plus, since it's a software, it's not like they can bring it back to the store to return it. Now, they've paid for a new service that they cannot get refunded and wasted time migrating information, and they have to spend even more time inputting the information back into the original software. They're right back where they started.

This choice reflects a risk-taking behavior where the immediate certainty of spending more is less appealing, even though it may lead to greater financial loss in the future should the product or service fail.

By embedding assurances throughout the digital journey, from secure checkout processes to clear return policies, companies can provide the peace of mind that answers the critical questions before they're even asked, ensuring that the customer's journey from contemplation to completion is as smooth and worry-free as possible.

To make a truly successful conversion, you want your customer to be confident in their purchasing decision.

Specifically, shoppers need these three things before they buy:

- Confidence their transaction will be secure
- Belief that your products are accurately described
- Assurance that you will deliver on your promises

Here are a few ways to build trust throughout the digital journey:[81]

- **Ease of operation**: Remember, perception is everything. Part of their perception is about usability—we tend to trust those who obviously know what they're doing—and part of that is psychological and subliminal. This means the entire digital journey from start to finish needs to run with as few stuck points as possible. That covers everything from fast site speed to easy-to-use filters, clear navigation, and

---

81 https://thegood.com/insights/website-credibility/

links that take you exactly where customers think it's going to go.

- **Audience knowledge:** How can customers trust you if they don't think you know who they are? We've seen companies hire designers who have great ideas but poor knowledge of the audience, leading to choices that don't reflect their customers at all. The results are invariably disastrous. You'll see few cowboys buying their jeans at Bergdorf Goodman. And you'll see few socialites purchasing dresses at Tractor Supply.

- **Contact information:** Customers want to know you are real. And real companies have physical addresses and phone numbers, not just email addresses.

- **Error-free pages:** If you can't be trusted to spell a common word correctly, how can shoppers trust you to get their order right? We've even seen educational websites with misspelled words. That's a prime example of a confidence sinkhole. The more capable you are of getting the simple things right, the more your customers will believe you can get the bigger things right.

DECISION-MAKING AND CONVERSIONS

- **Behind-the-scenes peeks:** People buy from those they know, like, and trust. While the "About" or "Meet the Team" pages are typically not a part of the conversion journey, their presence on the site outside of the sales-focused digital journey adds an extra layer of authenticity and authority to the overall process.

    Use these moments to share your company story or, better yet, show that there are real people behind the screen who work hard to make their purchase possible. Don't stop at headshots of your CEO and officers. Show behind-the-scenes shots of your customer service people helping solve problems, your production line getting orders ready, your developers working on the latest challenge. People like to do business with people, not with companies. The more personal you are, the more opportunity you have for building trust and confidence (a.k.a. website credibility).

- **Accreditations, awards, and affiliations:** It's one thing for a company to toot its own horn about the good it does, but it's quite another when that praise comes from outside sources. When customers aren't experts in what they're buying, they often turn to a company's reputation to help make their decisions. Clearly showcasing your affiliations, awards, and accreditations gives an added sense of authority that your company is backed by an unbiased, trusted source. We've consistently found trust badges to be an impactful approach to not only improving conversions, but also reducing cart abandonment rates.

In fact, this plays into another psychological principle—the noble edge effect, which shows that customers are drawn to companies that demonstrate a commitment to social responsibility.[82] But the key here is that it needs to be real. Otherwise, these efforts will backfire and take away trust. It's all about the heart behind a company's actions. When a business shows it cares through genuine social responsibility—like donating to causes or ensuring products are ethically made—customers tend to sit up and take notice.

- **Social proof:** Social proof is a huge persuader. Your visitors will believe what their peers say quicker than they'll accept the claims in your sales copy. Don't stop at posting customer reviews, though. You can also show logos and lists of news organizations who've written about you or snippets of headlines and stories that establish your abilities.

    Even negative reviews can add a layer of trust. Products and services with all five-star reviews look fake. When you do get a negative review, answer it sincerely and without defensiveness.

---

[82] Chernev, A., & Blair, S. (2015). Doing well by doing good: the benevolent halo of corporate social responsibility. *Journal of Consumer Research*, 41(6), 1412–1425. https://doi.org/10.1086/680089

Now that you've established trust, it's time to guarantee that you'll follow through with the promises you've made. You can do this through:

- **Money-back guarantees**: Shoppers want to know they can trust you. Indicating your willingness to stand behind your products is a great way to do that. The promise of a refund if the product fails to meet expectations, for example, reduces the perceived risk. Money-back guarantees can also serve as a testament to a company's customer service standards. It shows that a company prioritizes customer satisfaction and is prepared to go the extra mile to ensure a positive shopping experience. This level of service can differentiate a company in a crowded online marketplace and foster a sense of loyalty among customers.

- **Lifetime guarantees**: Nothing feels safer than knowing you're taken care of for the rest of your life. Or at least, your purchase is. Many companies offer lifetime guarantees but bury this benefit in the footnotes of their website. Building trust is as easy as pulling it to the front lines.

    As an example, when our team at The Good worked with a well-known premium camping brand, they already had a lifetime guarantee on all their gear, but the way it was buried in the website's footer under a vague "guarantees" button made it practically invisible. We took a different approach by prominently advertising this lifetime warranty throughout their category and product pages. We fine-tuned the wording to make it more appealing, calling it a "lifetime guarantee." Customers feel reassured that they can get their gear repaired if it ever gets damaged for the rest of their lives, creating a sense of low risk.

- **Hassle-free returns**: If your customers know they can simply return an item if it's not right, the decision becomes that much easier.

   Amazon, for instance, excels not only because of its lightning-fast two-day delivery, a point often emphasized, but also due to its seamless and free return process. If you want to make a return, all you need to do is walk to the Amazon return center at your local Whole Foods, Kohls, or UPS store and put your item on a counter. The worker will grab a box, place it inside, and scan a QR code on your phone. Before you even get back to your car, you'll have an email confirmation for the completed return and the money is credited back to your Amazon account.

In SaaS, thirty-day money-back guarantees give the customer a risk-free way to fully integrate into the product before deciding if it's right for them.

By prominently featuring your return policy on your product page, you can significantly boost conversions by reassuring potential customers that if the product isn't a perfect fit, they have a hassle-free way to rectify the situation.

Essentially, each step in the product journey provides an opportunity to strategically build trust between you and the customer. Whether it's how you list out the features and benefits, how you detail the product description, or even how you frame the product in the photos, each element plays a part in influencing the buyer's decision.

QUESTION:
HOW EASY IS IT FOR ME TO PURCHASE?

At this point, you're really getting practical. This is where you step outside the realm of how the customer feels into their actual, lived experience.

The conversion doesn't automatically happen when they've mentally committed to buying. If your customer has a product page open on their mobile browser for two weeks

DECISION-MAKING AND CONVERSIONS

without taking any action, they haven't technically converted, even if they've already made the decision to buy. The distinction lies in transitioning from the Information-Gathering and Decision-Making Phases to the stage where they take tangible action.

The act of making the payment is the final step in that process.

This is where many companies taper off on their journey. They assume that just because they got the customer to this decision point, just because they have the items in their cart, the conversion is inevitable. It's the "you can take it from here," approach.

But their abandon cart rates beg to differ.

While we love to see companies stand out from the crowd, your checkout page is not the time to flex your unique digital journey. As we've already touched on, customers need to have a certain level of trust to give up their valuable credit card information. When companies try to reinvent the checkout process, it can lead to confusion and loss of confidence from the consumer—and that's leaving money on the table.

When a customer reaches the checkout phase, they should not have to guess what to do next. The process should be so streamlined and intuitive that they can proceed without a second thought. If customers find themselves puzzled over how to add items to their cart, apply discount codes, or initiate the checkout process, it's a sign that the website design has strayed too far from the conventional path that customers are accustomed to.

Your job as a company is to make sure that the buying process isn't a problem your customers need to solve.

SOLUTION:
STREAMLINE THE CHECKOUT PROCESS

Optimizing your digital journey is one thing. But if those visitors never make it to the confirmation page, none of the rest really matters. The success of your company's website rises and falls on how many sales you make, not on how many people you can attract to your website.

According to research from Baymard, average cart abandonment hovers around 70 percent. At the end of 2020, one out of every five shoppers abandoned cart due to a "too long/complicated checkout process."[83]

To avoid this, you'll want to:

- **Eliminate distractions**. Because there are so many external factors at play, the longer it takes for your customer to find what they need, the more likely it is for a customer to become distracted. Anything that

---

83 https://thegood.com/insights/optimize-checkout-process

DECISION-MAKING AND CONVERSIONS

distracts them from the task at hand—hitting "pay now"—is going to hurt conversions.

This plays into a psychological principle called attentional bias.[84] Essentially, it's our natural tendency to home in on certain things while overlooking others. So if the path to purchase is too complex or there's just too much going on, customers might get sidetracked by details that aren't crucial to their buying decision.

---

84 Julian, K., Beard, C., Schmidt, N. B., Powers, M. B., & Smits, J. A. (2012). Attention training to reduce attention bias and social stressor reactivity: An attempt to replicate and extend previous findings. *Behaviour Research and Therapy*, 50(5), 350-358. https://doi.org/10.1016/j.brat.2012.02.015

A clear and simple process, free from distractions, helps guide customers right to the finish line. It's about making sure that the only things catching their eye are the things that help them check out smoothly.

Shopify, for example, excels in preventing users from making poor design choices—they've standardized their checkout design to such an extent that it's widely trusted.

- **Calculate shipping and/or taxes before you ask for payment details.** It's amazing how such a small sequencing detail can make such a big difference in conversions. In our work at The Good, we consistently highlight shipping and tax costs as the primary reason people abandon their shopping carts.

  It's astonishing how even a relatively small fee, like $4 on an otherwise pricey product, can deter customers when it's sprung on them as a surprise. It leaves them feeling deceived and reluctant to proceed. State all added costs on the first page of your cart. Hiding this until the last step of checkout will certainly frustrate buyers and lose the sale.

- **Don't require shoppers to register before checkout.** Registration is a big source of friction on websites.

DECISION-MAKING AND CONVERSIONS

Offer a guest checkout procedure so your users enjoy a smoother experience. And ideally only offer registration once they've completed the order.

- **Make your return policy prominent and matter-of-fact to reduce fear and build trust.** An unsatisfactory or hidden return policy can stop you from getting the order. Online shoppers often count on a photo and description to evaluate a product. They want to feel assured they'll be satisfied with the item when it arrives—whether that's at your home or in your inbox. Your return policy can help remove any fear they have about placing the order.[85]

- **Provide clear action items.** Make "add to cart" the most obvious next action on every product detail page. Your path to purchase should be clear and simple.

- **Incorporate a sticky/quick add to cart button.** The sticky "add to cart" button is a design element e-commerce companies use to keep the "add to cart" button visible on a mobile screen while customers are browsing. It can hover near the top or the bottom of the

---

85 https://thegood.com/insights/reduce-ecommerce-shopping-cart-abandonment-today

screen to give shoppers the chance to tap it wherever they are on the page. This can bring substantial conversion boosts with both desktop and mobile customers.[86]

- **Leverage one-click shopping for repeat customers to make it even easier for them to keep coming back and to avoid cart abandonment.** Shopping online is no longer a novelty or something you do because a certain item isn't available locally. For many people, ordering online is now the preferred choice. No driving, no fighting crowds, no standing in line. When you make shopping with you as easy as loading a basket and clicking a "buy now" button, shopper loyalty will grow.

- **Make it easy for shoppers to make changes to the cart in order to prevent frustration and cart abandonment and keep them engaged.** This part of checkout can be especially frustrating. You get to the end of the procedure and realize you forgot something. Now what? The optimized procedure will make changes to the cart easy to do.

---

86 https://thegood.com/insights/sticky-add-to-cart/

DECISION-MAKING AND CONVERSIONS

- **Optimize your in-cart add-ons.** We know we've looked at the digital journey thus far as a series of carefully thought-out decisions, but there's still room for impulse purchases along the way. In-cart add-ons, especially those that are personalized to this specific customer based on their current purchases or purchase history, are a great way to capitalize on the decision-making momentum you've built up until this point.

Let's See This in Action

At checkout, timing is everything.

An upsell in the right place can bring huge conversions. But placed wrong, and it might cost you the sale the customer came for in the first place.

While optimizing a client's checkout process, we found a full page of additional cart add-ons in the middle of checkout. The idea was for people to add more to their cart before submitting their payment. But in reality, it just frustrated customers who were suddenly yanked out of the flow of checkout.

Why? Because these add-ons weren't in the right place. Once a user goes past their cart, they've already made their decision. So if you're promoting an add-on, you want to give them the options before they finalize their choices, not after.

Showcasing upsells in the cart before checkout gave customers the opportunity to explore new options before finalizing their purchasing decision.

Because they saw it at the right time, they were willing to convert more—bringing in $328,000 in annual revenue.

- **Avoid surprises.** Doing things like not telling the buyer the product they want is out of stock until they've ordered the item are surefire ways to create animosity. Never surprise your customers with bad news.

- **Provide estimated delivery dates.** Ordering a product without knowing when you'll receive it is almost like making a blind purchase, and most customers won't go for that. They want clarity on delivery times. Estimated delivery dates (EDD) predict when a product will arrive at the customer's specified address.

In fact, three-fourths of shoppers indicated that putting an EDD on the product page or in the cart positively influences their decision to buy an item.[87] This goes back to our psychological desire for certainty whenever possible. Studies have even shown that what users care about isn't "shipping speed" but rather the specific date of delivery, like "When will I receive my order?"[88]

- **Offer delivery guarantees.** Delivery guarantees are another aspect to consider, especially for time-sensitive products or events. For example, in situations where customers are relying on a product to arrive on a specific date, offering a time-based guarantee can instill confidence. However, it's crucial to honor such guarantees—failing to meet promised delivery times can erode trust.

Of all the aspects of your website, the checkout page should be the one your company tests and optimizes the most

---

[87] https://www.getconvey.com/blog-b-consumer-research-estimated-delivery-date/
[88] https://baymard.com/

frequently. It's important to audit the entire checkout procedure to maximize conversions. Spot checks of stuck points along the path to purchase are typically required when an obvious problem is discovered. Regular full-system inspections of the entire checkout process can reveal those revenue-draining areas before they show up on the radar.

AT A GLANCE

In the Decision-Making and Conversion Phase, Your Customer Will Ask

- Am I ready to make a purchase?
    - **Solution 1:** Get customers ready to convert with an emotional appeal.
    - **Solution 2:** Hit them with a meaningful and well-timed call to action.
    - **Solution 3:** Make customers feel like they already own it.

- What should I choose?
    - **Solution 1:** Remove all barriers that make the choice more complicated.
    - **Solution 2:** Compare wisely.

DECISION-MAKING AND CONVERSIONS

- What's in it for me?
  - **Solution:** Don't discount—add value.

- If I'm wrong about this choice, what is the worst possible outcome I could experience?
  - **Solution:** Build trust and then guarantee it.

- How easy is it for me to purchase?
  - **Solution:** Streamline the checkout process.

Principles at Play

- **Action bias:** When given the option, people would rather take action instead of doing nothing.

- **Priming effect:** The tendency to make connections and decisions based on the subtle cues around us.

- **Affect heuristic:** The idea that people put more stock into emotional appeals than logic.

- **"Bye-now" effect:** Merely reading the word "bye" nudges us to think of "buy."

- **Ikea effect:** People feel more attached to items they've created themselves.

- **Decision fatigue:** The more decisions we make in a day, the more mentally drained we feel, and the worse decisions we make.

- **Less-is-better effect:** Shoppers may actually favor an item of lower quality when it's the only one they're considering, as opposed to comparing it alongside better options.

- **Take-the-best heuristic:** Instead of carefully considering all elements to the decision, customers may make their decision on one key factor.

- **Decoy effect:** The presence of the decoy simplifies the decision-making process by providing an easy-to-reject option, making the target seem more appealing and valuable in comparison

- **Incentivization:** By offering rewards such as discounts or special bonuses, companies activate the reward-seeking circuits in our brains, leading us to engage in behaviors that could earn us those rewards.

- **Negative intent shaming:** The practice of a company attempting to dissuade a consumer from taking a particular action the company doesn't want them to take.

- **Motivating uncertainty effect**: The excitement associated with rewards of uncertain value often drives us more than the promise of certain rewards.

- **Bundling bias**: A cognitive bias that affects how we perceive the value of bundled packages and gift cards.

- **Loss aversion**: The discomfort we feel from a loss is more intense than the joy of an equivalent gain.

- **Status quo bias**: A natural preference for the familiar, partly because it's seen as less risky than the unknown.

- **Noble edge effect**: Customers are drawn to companies that demonstrate a commitment to social responsibility.

- **Attentional bias**: The natural tendency to home in on certain things while overlooking others.

Optimization Opportunities

- Call to action
- Personalization
- Product or service comparisons
- Checkout

- Pricing pages
- Company promotions

In Summary

- Customers want to take action. So if your target audience isn't converting, it means your digital journey includes a stuck point that gets in the way of achieving their goal.

- Customers trust their gut when making decisions. How they feel about the choice is more important than the choice itself.

- Evoking emotions throughout the digital experience can lead customers toward the right decision.

- Often, the catalyst from the Information-Gathering Phase to the Decision-Making and Conversion Phase lies in a strong call to action.

- Choice overload in the decision-making process can make customers abandon the digital experience altogether.

- Customers will never be left with a difficult choice if, by this stage of the digital journey, you've properly guided

DECISION-MAKING AND CONVERSIONS

them toward the best option (or drastically reduced set of options) for their needs.

- The more choices customers have, the harder it is to make a decision, and the less confident they are in their decision.

- The way you present options for comparison can impact the customer's choice.

- Using incentives to entice customers to buy might lead to short-term gains, but they could also decrease the margin for error and consumer patience. Instead, add value.

- If you want customers to convert, you need to earn their trust first.

- A clear and easy checkout process is essential in getting customers to convert.

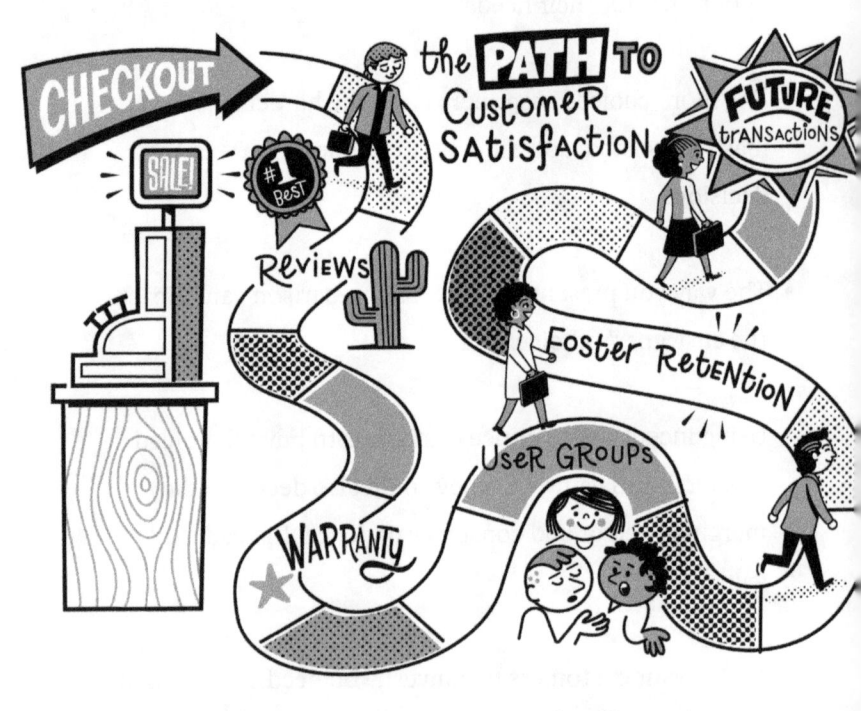

Despite a smooth journey leading to a successful checkout, the journey doesn't end there. A complete digital experience not only guarantees customer satisfaction but also fosters retention and future transactions.

5

POST-PURCHASE

ONVERSIONS DON'T STOP AT CHECKOUT.

By now, you may have curated an expertly guided customer digital journey. You may have ushered customers along toward the right decision as easily and efficiently as possible—in ways that work with their cognitive shortcuts. You may have led them to a successful checkout. The purchase confirmation may be in their email. But their experience doesn't end there.

A great digital experience follows all the way through—which means long after the product or service is delivered. The post-purchase experience, from the confirmation email to unboxing or platform log-in, can either validate the customer's choice or trigger regret.

In the Information-Gathering Phase, we talked about the recency effect—the idea that we are most likely to remember the last thing we saw. In that phase, we talked about it in terms of sequencing elements of your website like your navigation. But now, we're applying it to the experience as a whole. The last impression a customer has of your company is the one that sticks.

In other words, you could have the most flawless digital journey on the internet, but if the post-purchase experience doesn't meet expectations, it's all for nothing.

If you do it right, this is the stage that turns one-time buyers into repeat customers, that turns negative reviews into lifelong company advocates, that keeps your return rate low and customer sentiment high.

But for some reason, this is where company teams tend to stop actively planning the experience—they simply show a Shopify checkout confirmation and think the job is done. However, there's room for optimization at this final stage to significantly enhance the customer's perception of the company so that their last impression is positive (even if it didn't start that way).

Many companies are solely focused on getting that initial conversion and neglect the lifetime value of the customer. This oversight can be costly, not only in terms of lost revenue

POST-PURCHASE

but also in increased returns due to buyer's remorse or other post-purchase challenges. Companies have the opportunity to solve these problems at that final point of contact, making the customer feel well taken care of and well informed about the next steps, delivery times, onboarding process, and so on.

Here, you're not focusing on conversions—you're focusing on increasing customer sentiment. This, in turn, will attract even more conversions. Done right, it's an ever growing sales cycle. But ignored, it stops the conversion at one. Or worse—at a return.

For better or worse, we often make purchases—or decline to purchase—based on our feelings and emotions. If we don't like a company, we won't buy from them, even if their products and services check off all of our other boxes.[89]

Customer sentiment refers to the emotions customers feel toward your company, products, or services. It helps you understand whether your customers' feelings are positive, negative, or neutral, as well as why they feel that way.

Generally speaking, if your customers have positive sentiment, they are more likely to buy and become repeat loyal customers. Customers who think of your company negatively are less likely to buy.

To formulate these opinions, customers may ask themselves:

---

89   https://thegood.com/insights/customer-sentiment

- Were my expectations met?
- Would I make this decision again or recommend that a friend do the same?

QUESTION: WERE MY EXPECTATIONS MET?

Remember, to get the customer to the point of purchase, you likely made a lot of promises along the way. Now, it's time to follow through.

You as a company should do everything in your power to fulfill every promise you've made along the customer journey, but the reality is, life exists. Things happen. Sometimes a delivery truck that has nothing to do with your company gets a flat tire on their route, and their package comes a day late. Sometimes supply chains break down and cause a backlog of delayed shipments that you had no way of foreseeing. It happens.

It's not necessarily the company's fault—but try telling the customer that.

While your company can help boost customer sentiment using psychological principles, there's one principle that can work against you—and there's not much you can do about it.

This is called the fundamental attribution error, which states that when forming opinions, people tend to give excessive weight to personal (or company) traits while minimizing the impact of their surrounding circumstances.[90]

Sometimes your customer sentiment is out of your control. They may not account for logical factors (like shipping delays caused by a storm) and instead place blame on your company. Effectively managing the impact of the fundamental attribution error requires a thoughtful and strategic approach.

It's crucial to ensure that the company's values and promises align seamlessly with the actual customer experience as often as possible. When there's a disconnect between what the company pledges and what customers experience, it can create a perception that any negative occurrences are solely the company's fault, eroding trust and overall sentiment.

And when you can't follow through on the promises you made, you jump into action to make it right.

SOLUTION:
PROACTIVE COMMUNICATION FOR THE GOOD *AND* THE BAD

In the post-purchase journey, proactive communication is key, particularly when unexpected issues arise. Rather than

---

[90] Chater, N., & Loewenstein, G. (2016). The under-appreciated drive for sense-making. *Journal of Economic Behavior & Organization*, 126, 137-154.

assuming that customers will naturally understand situational factors, you step in with clear explanations and effective solutions. Addressing customer concerns transparently and demonstrating empathy can help counteract the negative effects of this bias, showing customers that your company is dedicated to their satisfaction.

This means if you've ever had the urge to delete a negative review—don't.

As we've covered earlier in the digital journey, social proof, especially in the form of reviews, testimonials, social media mentions, or company ambassadors, is one of the best ways to show customers that they can trust you, your product and your company.[91]

At this stage, it's time to focus on social proof in the form of reviews. This can be a two-headed monster. Allowing potential customers to see reviews can increase your conversion rate. But it feels like allowing negative product reviews on your website risks turning buyers away.

Allowing negative reviews and appropriately responding to them gives your company the opportunity to:

- Fix or address a one-off issue that doesn't represent the typical digital experience

---

91   https://thegood.com/insights/product-reviews-improve-conversion-rates

POST-PURCHASE

- Understand common issues with your product or service so you can fix them
- Add credibility to your positive reviews—no one trusts a company with only five-star reviews

Actually, the fact that the customer left a review at all shows an opportunity for improvement. Once customers leave reviews, even neutral or negative ones, they tend to feel more favorable toward the company. They feel as if they've contributed to the company, drawing them closer to it.

The first step to utilizing these reviews—good or bad—is to get them in the first place. If you want customers to write reviews, the first step is pretty obvious: ask them to do it.

You can prime shoppers for reviews by letting your customers know three things immediately after the transaction is completed:

- You want them to be 100 percent satisfied with their purchases.
- If they aren't satisfied, you want to know so you can correct the problem.
- Once they are completely satisfied, you would appreciate an honest review.

In fact, you can take the stars out completely. Instead of letting a review stop at a star rating, ask the shopper for useful information. For example, how did the product fit? And was it true to size? How was the onboarding process? Is the platform easy to navigate? Then turn that data into a tool to help other customers.

When the negative reviews roll in—because let's face it, even the most successful companies in the world aren't perfect—you can see them as an opportunity, not an obstacle. Here's where to start:

- **Learn from them.** Criticism can be a valuable teacher. Assess and improve upon your product. Examine what customers are not satisfied with and how you can start to improve on that. Negative reviews, believe it or not, can actually end up benefiting your business.
  - *Note*: It's important to only approve negative reviews from verified buyers. If you don't have a system for verifying reviewers, what's stopping your competition from going on your website and leaving a series of negative reviews?

- **Respond appropriately to negative reviews.** This doesn't mean that you have to offer a full refund in every case. It simply means that you need to express real interest in the problem and put a good-faith effort into fixing it. You want to show customers that you genuinely do stand behind your products and take problems seriously.

- **Respond with gratitude to positive reviews.** If a customer takes the time and effort to leave you a positive review, you should take the time to acknowledge it and express your appreciation. If someone said to you in person, "I love your product/service, and I want to tell everyone about it," wouldn't you thank that person? Treat your online customers in the same way.

> ### Let's See This in Action
>
> An e-commerce clothing company working with our team was receiving customer reviews, but after user testing and research, we hypothesized that adding fit based on customer reviews would give customers more confidence when purchasing a product.
>
> So we ran a test that added a size and fit scale based on customer ratings, and it was a winner.
>
> Based on the lift in per-session value, implementing the winning variant to all users would produce a lift in annualized revenue of over $276,000.
>
> In this case, using the reviews to help customers gain confidence in the size and fit of their purchase

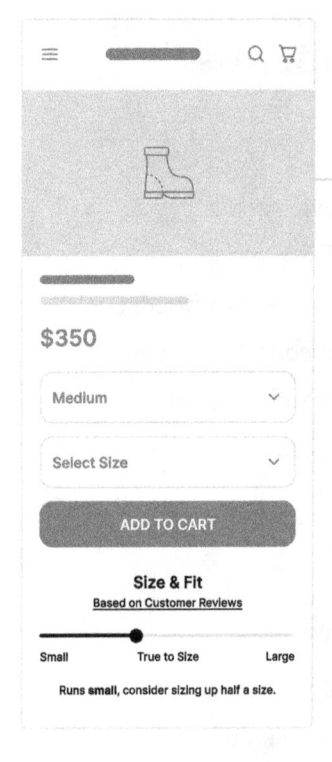

both improved the user experience and increased transactions.

Reviews either validate that the digital journey was a good one or give your company an opportunity to turn a bad journey into a better one.

A customer who had an issue that was successfully resolved tends to become a stronger advocate than someone who never faced any problems at all. Let's say you received the wrong product, encountered an issue, or simply didn't like the item, but the return process was seamless, and the company handled it well. You're much more likely to become a loyal advocate.

QUESTION
DID THIS PRODUCT OR SERVICE WORK AS IT SHOULD?

In the Post-Purchase Phase, you aren't guiding customers toward a purchase anymore—you're ensuring that they continue to feel confident in the purchase they did make. The key to this is ensuring that the product or service they received is exactly what they thought it was going to be. The only good surprise after that is for a customer to find out that the product or service is even *better* than they expected.

For e-commerce, this comes down to quality. Customers want to see that you live up to every claim you promised along the digital journey when they finally have that item in their hands.

For SaaS, it's a completely different experience, and one we are going to focus on primarily in this question. The moment customers download your software, log into your platform, or start using your app, they're embarking on a whole new digital journey.

You expertly led them here, and you can't abandon them now.

SOLUTION:
CREATE A CLEAR AND EASY ONBOARDING PROCESS

Your responsibility to guide customers extends well into the Post-Purchase Phase.

The service, platform, or app needs to fulfill every feature and function that prompted the customer's decision. You may live up to every promise you've made, but if the customer doesn't know where to access it or how to use it, it's as if those features don't exist at all.

If the customer feels like they don't exist—whether that's true or not—you aren't fulfilling their expectations.

This is where the onboarding process becomes crucial. Most onboarding experiences can make or break a product.

With so much competition, you can't afford to confuse users even for a second. In fact, 74 percent of potential customers will switch to another solution if the onboarding process is complicated, while 86 percent will stick around for the long haul if they have an enjoyable onboarding experience and get continuous education.[92]

Your goal might be to get someone to click the "sign up" button, but what then? Making sure the next steps are easy and insightful can do wonders for your business. The user onboarding process is often the first point of contact you have with new users, which means it's your one chance to make a good first impression (remember how important that is?).

Get it right, and it can lead to increased usage rate, higher retention, and long-term customer satisfaction.

A really good onboarding experience can:

- Increase retention by helping new users quickly understand the value of your product

---

[92] https://userpilot.com/blog/saas-user-onboarding-2021

- Improve customer satisfaction by reducing friction and confusion
- Increase conversion rates by converting free trial users into paying customers
- Improve product feedback by gaining insight into how users interact with your product

An excellent onboarding process is going to vary based on user needs and digital savviness. A beginner-friendly platform is going to need a much more intuitive and streamlined process than an extremely niche software aimed at advanced professionals, where there's a little more room for complex options and pathways.

That being said, while we can't map out an exact blueprint for every single user onboarding flow, there are some commonalities to bear in mind:

- **Put the user first.** The best user onboarding experiences focus on the needs of each individual user and what they need to know to get started. It might be tempting to show off your awesome product features, but now's the time to be humble and empathetic.

- **Promote user action.** Quick wins are a must in any onboarding experience. Focus on getting new sign-ups to complete small actions quickly so they start to feel comfortable using the tool ASAP.

- **Push the value.** Your onboarding experience should highlight the value of your product and why it's different from similar ones out there. Ideally, you want to help the user gain value as soon as possible—this will ensure they keep coming back.

- **Collect feedback.** A good onboarding experience should meet the needs of the user. Track and measure how successful your process is by seeing how many people stick around and asking new sign-ups for their feedback. You can use this to tweak and improve the process.

- **Take it slowly.** The last thing you want to do is information dump everything onto new sign-ups. Avoid overwhelming by trickling out information and allowing users to go at their own pace.

The way you present these ideals as an onboarding experience can vary widely. Here are a few options that work well for some of the most user-friendly SaaS companies out there.

POST-PURCHASE

Choose Your Own Adventure

Tailoring your user onboarding flow to the unique needs of each user will increase loyalty—and usage. No one wants to walk through a demo that's not relevant to them.

Instead, put users in charge of their own destiny. Not only does this create a highly relevant process for new sign-ups, but it also gives you more information about how people are using your product.

For example, Canva navigates this perfectly by asking new sign-ups what they plan to use the tool for. The answer they choose then guides the rest of the onboarding experience.

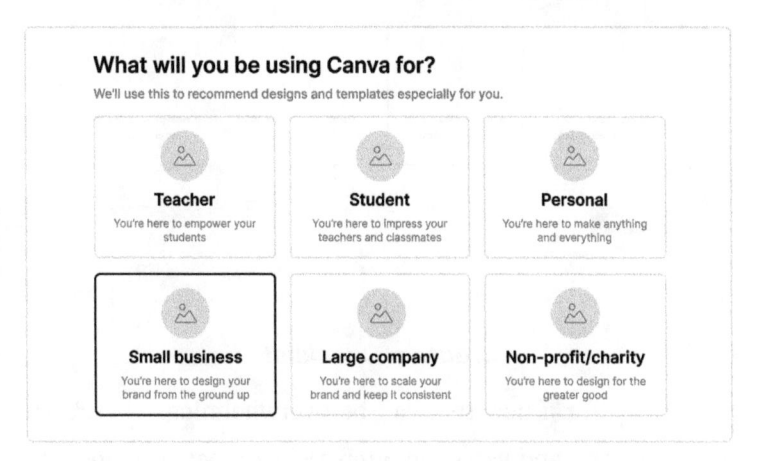

What will you be using Canva for?
We'll use this to recommend designs and templates especially for you.

Teacher	Student	Personal
You're here to empower your students	You're here to impress your teachers and classmates	You're here to make anything and everything
Small business	Large company	Non-profit/charity
You're here to design your brand from the ground up	You're here to scale your brand and keep it consistent	You're here to design for the greater good

Personalization Quizzes

Someone has just chosen to use your product over the thousands of others out there—that's huge. Now it's time to make them feel special.

Get to know your users, and help them get started as quickly as possible. This sets them up for a series of quick wins that will increase the chances of ongoing satisfaction and loyalty.

For example, Asana does this by asking a handful of questions about the new user and then helping them set up a new project based on their actual needs. This means new sign-ups are ready to hit the ground running as soon as they finish onboarding.

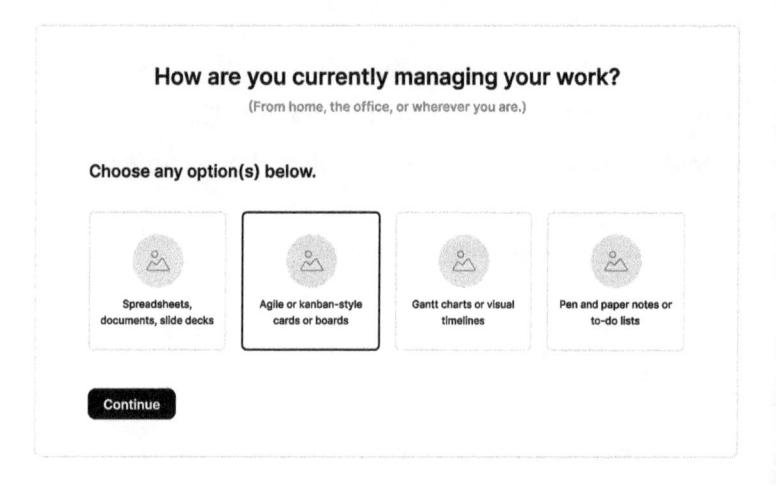

Create a Checklist

Signing up for a new tool can be overwhelming. There's so much to learn, and users can often feel like they're a long way from where they need to be when they get started. To make it easier, create a checklist that users can work through at their own pace. Make sure each item provides a quick win so that new sign-ups feel confident moving on to the next step.

HubSpot, for example, reduces the overwhelm by providing a very clear user checklist. New sign-ups can work their way through the checklist at their own pace, but they can always see what's up next and how many more items there are to tick off.

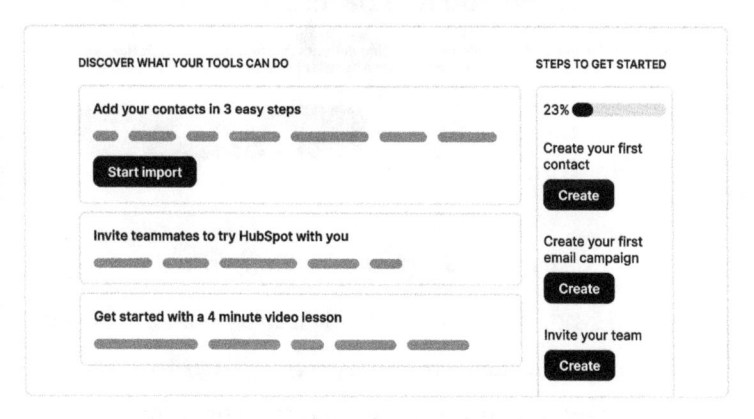

DISCOVER WHAT YOUR TOOLS CAN DO	STEPS TO GET STARTED
Add your contacts in 3 easy steps Start import	23% Create your first contact Create
Invite teammates to try HubSpot with you	Create your first email campaign Create
Get started with a 4 minute video lesson	Invite your team Create

Take Customers on a Tour

Seventy-five percent of people believe video is the best way to learn how a product works.[93] But dumping all the information a new user needs to know in one long video is a surefire way to overwhelm—who really wants to sit through an hourlong video that may or may not even be relevant to their needs?

Instead, give new users small chunks of information, and provide positive reinforcement every time they take the next step.

Toggl addresses this by implementing a series of bite-size videos that cover each feature. At the end of the video series,

---

[93] https://www.wyzowl.com/app-demo-statistics/

users get positive reinforcement from Toggl and encouragement to continue using the tool.

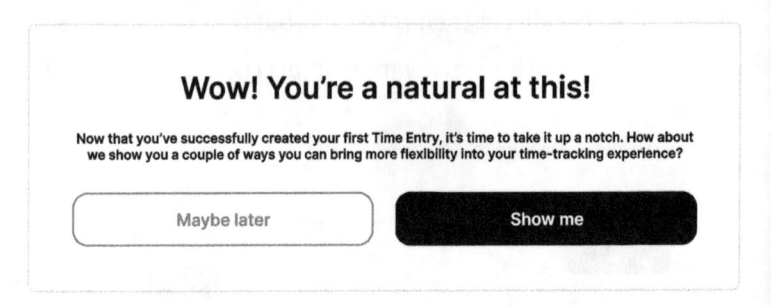

> ### Wow! You're a natural at this!
>
> Now that you've successfully created your first Time Entry, it's time to take it up a notch. How about we show you a couple of ways you can bring more flexibility into your time-tracking experience?
>
> [Maybe later]   [**Show me**]

Get Users Started with Templates

Blank page syndrome is real. Provide a collection of templates that not only show users how they can use your tool, but that also allow new sign-ups to get started straight away.

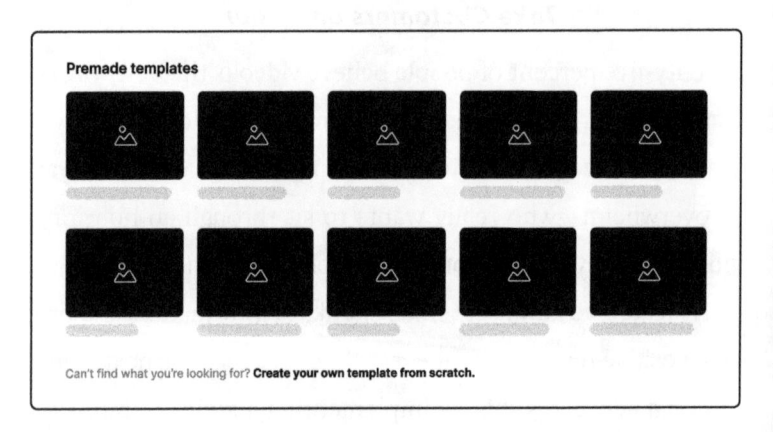

Premade templates

Can't find what you're looking for? **Create your own template from scratch.**

Process Place avoids putting its new users through this by showcasing a handful of templates they can get started with. Users can browse a library of ready-made templates they can edit and use straight away without figuring out how to use the tool from scratch.

Create an Interactive Walk-Through

Don't try to cram every use case into your onboarding process. Instead, give new sign-ups the option to choose which product features are most important to them, and encourage them to take action with interactive to-dos.

Notion has such a mind-boggling number of use cases it can be tricky to condense all the possibilities into one simple onboarding process. To tackle this, the tool has created an interactive product walk-through that lets users pick and choose what information they want to learn first.

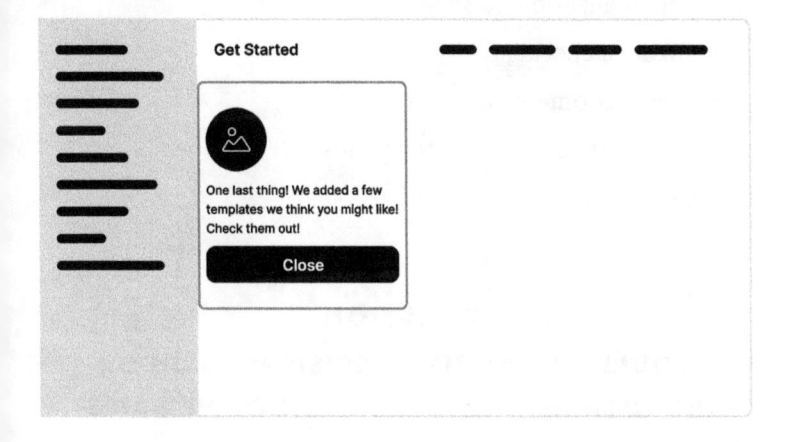

Building a user-centric onboarding experience is only one piece of the puzzle. Once a user is signed up (or subscribed), your job is to discourage cancellation by delivering a seamless experience that continues to surprise and delight.

A positive onboarding experience should take the individual needs of each new sign-up into account. It should get them to their "aha!" moment as quickly as possible and give them the chance to explore your product at their own pace.

Putting It All Together

Great SaaS experiences hinge on successful onboarding. That's going to look different for every company. Here are some tactics you can try:

- Choose your own adventure
- Personalization quizzes
- Create a checklist
- Take customers on a tour
- Get users started with templates
- Create an interactive walk-through

QUESTION:
WOULD I MAKE THIS DECISION AGAIN OR RECOMMEND THAT A FRIEND DO THE SAME?

This question is 10 percent "Does this product actually fit my need enough to want it again or recommend it to a friend?"

and 90 percent "Does this purchase *feel* like it was worth it?" And honestly, those are often two different answers.

Many companies miss out on retention, referrals, and repeat business because they offer services that precisely meet the customer's needs, leading to a satisfying purchase. However, they often fail to reengage these customers for future needs. The issue isn't a lack of customer interest in repurchasing; it's that the company successfully attracts the customer but then releases them too easily after the sale.

They close the sales loop so completely that it becomes easy for customers to move on.

The key is to leave an open door for customers, encouraging them to return. This way, when they do come back, they don't feel like they're starting from scratch. Instead, they are welcomed back to continue their journey, reassured by their previous positive experience and confident that the company will again fulfill its promise.

SOLUTION:
LEAVE THEM ON A GOOD NOTE WITH POST-PURCHASE EMAILS

If you structured your digital journey correctly—from Information-Gathering all the way through to Post-Purchase—the positive experience should speak for itself, right?

Maybe. Maybe not. Let's not leave it up to chance. That's where post-purchase emails come in.

Now, you're not focusing on bringing in new customers—you're making sure the ones you attract stick around.

There is only a 5 to 20 percent chance of a new consumer buying from you.[94] In contrast, loyal customers are five times more likely to repurchase.[95] And since it is seven times more expensive to acquire a new customer than to retain customer relationships, you probably don't want to mess this up.

This sequence should include:

- An order confirmation that tells the customer:
    - Their order went through
    - When it will ship and how to track their shipment
    - How they can access their receipt
    - Who to contact if there's a problem
    - What they should expect next

- A shipping confirmation with additional trust-building information such as:
    - A way to contact the customer service team

---

[94] https://www.outboundengine.com/blog/customer-retention-marketing-vs-customer-acquisition-marketing

[95] https://thegood.com/insights/5-post-purchase-emails

POST-PURCHASE

- Links to important information such as returns and refunds, FAQs, delivery time, and payment
- A recommended section to promote other items
- A CTA

- A check-in on the shipment that asks:
  - Did it arrive on time?
  - Did it arrive in good shape?
  - Did you have any issues after it got there?

- A request for a review, which can also include:
  - A reminder to customers about the great product or experience you provided, usually using an image
  - Honest and upfront information about how long the review or survey will take
  - A description of how the review or results of the survey will benefit the customer

Barring any additional unexpected issues, this request for review is the last point of contact you have with your customer, so make sure it's a good one.

Because now, it's time to ask for something in return.

If you want good feedback, good reviews, and repeat business, it's on you to hand them the opportunity to give you that.

The way you handle this follow-up can drastically influence how they respond going forward. By framing the request a certain way, companies can influence how a customer reacts—such as calls for reviews that ask you to "tell us what you love" (positive) versus "give feedback" (neutral) or "tell us how to improve" (negative).

This is due to the observer expectancy effect, which sheds light on how people's behavior can change when they are aware that they are being observed.[96]

---

[96] Nichols, A. L., Maner, J.K. (2008). The Good-Subject Effect: Investigating Participant Demand Characteristics. *The Journal of General Psychology.* 135(2), 151-165.

POST-PURCHASE

Understanding the observer expectancy effect can be valuable, especially when looking for feedback. While these common calls-to-action may seem like they're asking the same thing, they're going to get drastically different responses.

For example, asking, "Which features do you want to keep?" prompts the user to think about the benefits, whereas asking, "Which features do you want to cancel?" leads them to contemplate drawbacks. Instead of a generic "Contact us with any issues," you might say, "Feel free to share your thoughts, whether they're positive, negative, or neutral." This nudges people to provide feedback and assures them that all types of input are welcome.

So let's say you want feedback on a new coffee machine. Here's what you might get back:

- **What you say**: "Did you find our product difficult to use?" (Tone: negative)
  What the customer subconsciously thinks: "This product can be challenging, and now I should look for the flaws."
  The review: "I recently purchased this coffee machine, and I have to say, it was quite a disappointment. The machine's setup process was far from straightforward, and the instructions provided didn't offer much clarity. It took me a considerable amount of time and

frustration to figure out how to use it properly. The coffee is great, but the process it takes to get there isn't worth it. Plus, it's way too loud. It's a shame because I had heard great things about this machine, but it didn't live up to the hype."

- **What you say**: "Give us your feedback." (Tone: neutral)
  What the customer subconsciously thinks: "This company genuinely wants to know what I think, and my opinions are valued regardless of whether they are positive or negative. I should be honest and fair."
  The review: "The machine itself has been a reliable addition to my kitchen. It's easy to use once it's set up, and the coffee it brews is consistently good. It's a little louder than other models I've used (which isn't my favorite thing first thing in the morning), but once you get used to it, it's not so bad. Otherwise, I haven't encountered any major issues, and it has become a daily essential for me."

- **What you say**: "Tell us what you love about this experience." (Tone: positive)
  What the customer subconsciously thinks: "I really did love this—let's focus on all the positives."

The review: "I love this machine! The coffee it brews is simply fantastic. It's rich, aromatic, and consistently delicious. It's easy to operate, and the options for customization are a coffee lover's dream. Whether I'm in the mood for a bold espresso or a creamy latte, this machine delivers every time."

Remember, every question sets a direction for the respondent's thought process. When you see the subtext behind the words, you have the ability to subtly guide the user's decision-making process.

These are more than just a call to action or a request for feedback—they're influential tools that can shape behavior and decisions. Use them wisely.

AT A GLANCE

In the Post-Purchase Phase, Your Customer Will Ask:

- Were my expectations met?
    - **Solution:** Ensure proactive communication for the good and the bad.

- Did this product or service work as it should?
    - **Solution:** Create a clear and easy onboarding process.

- Would I make this decision again or recommend that a friend do the same?
  - **Solution:** Leave them on a good note with post-purchase emails.

Principles at Play

- **Recency effect:** The idea that we are most likely to remember the last thing we saw.

- **Fundamental Attribution error:** When forming opinions, people tend to give excessive weight to personal (or company) traits while minimizing the impact of their surrounding circumstances.

- **Observer Expectancy Effect:** People's behavior can change when they are aware that they are being observed.

Optimization Opportunities

- Social proof
- Onboarding
- Email follow-ups

In Summary

- The digital journey doesn't end at checkout. A complete digital experience sees the customer through post-purchase.

- In SaaS, your onboarding process can make or break your digital experience.

- The last impression you leave on your customers plays a vital role in their purchase satisfaction and lifetime value.

- Addressing customer concerns transparently and demonstrating empathy shows customers that you are dedicated to their satisfaction.

- Post-purchase follow-up opens the door for future purchases.

CONCLUSION

Digital journey optimization is filled with tactics and best practices—many of which we pioneered at The Good. Yes, exploring different types of A/B tests and diving into your company's data can lead to informed changes that bring big revenue results. That process merely scratches the surface of the opportunities you have to create a truly curated and guided digital experience.

By knowing the why behind these successes—the manner in which people think, the reasons they react to certain stimuli while ignoring others, and the rationale behind their decisions—you have the opportunity to present a digital journey that aligns with the way your customers think and feel, even if they don't understand why.

Digital optimization is not merely about seeing customers as data points but recognizing them as real humans with unique thought processes and behaviors. When you align your digital experience with their mental shortcuts and expectations, you make it easier for them to navigate and make decisions. Done correctly, this leads customers to the effortless, intuitive conclusion that your company offers the product or service they need, making their choice to purchase almost automatic.

Remember, the goal is to create a digital experience that not only meets but anticipates your customers' wants and needs.

This approach marks the transition from viewing your customers as mere collections of data to recognizing them as real,

CONCLUSION

complex human beings with unique thoughts, emotions, and behaviors. This is the turning point where your company prioritizes creating a path that caters to your customers' needs above all else.

No matter what else changes in the future, this focus on the customer experience will remain constant. Technology is going to change. Marketplaces are going to shift. But the psychological principles that influence how we make decisions will stay the same.

That's because these psychological principles have been around forever. Remember, many of them range all the way back to the cavemen—in-group bias, for example, can be traced back to the idea that people who were part of a larger group were more likely to survive.

There's a reason that these principles are so ingrained in our minds that we often don't even notice they are there—and that's because they're a part of human nature. These foundational psychological elements are just as true today as they were back when the cavemen used them to survive, and they'll be just as true twenty, fifty, or a hundred years from now.

It's the outcome of those principles that change.

Because as times evolve, so do people. Every day, people gather new experiences to draw upon later to make decisions. So when you think about the future of digital journey optimization, of course you'll need to study the changing trends around technology. A decade ago, for example, no one was buying on their phones. Now, it's one of the top ways customers purchase products. But if you look at that shift and say, "We need to optimize because technology changed," you're missing half of the story.

That shift to mobile shopping didn't occur just because it was more accessible—smartphones had been around for a long, long time before any company felt the need to optimize for mobile. That shift occurred when it did due to changes in the way people live. After the pandemic, people became much more comfortable using technology in ways they hadn't before. We weren't as tied to one physical location anymore as companies went remote. So we turned to our phones to get things done on the go.

The way we live our lives changed. The principles that led to the behavior didn't.

At the end of the day, those unconscious cognitive shortcuts fired away with one goal: to make the purchasing process as easy and efficient as possible. Before this shift, that meant heading to checkout on a desktop, where the photos are bigger, information is easier to read, and it's easier to input information without errors. But now, people are away from their computers more than they're on them, and Apple Pay and Google Pay make checkouts just a few clicks away. Same goal, different process.

It's vital that your website adapts accordingly.

CONCLUSION

And it's not enough to do this process once and say the digital experience is fixed. Optimization is not a destination but a continual journey. To create and maintain a seamless digital journey, companies should continuously monitor, test, and optimize every element of their website.

By understanding the psychological principles behind what works and what doesn't, you will have a much better understanding of what needs to be fixed when disruptions arise along the digital journey, regardless of what that digital journey looks like. With that deeper understanding, you can strategize targeted solutions instead of blindly applying tactics to see which one will stick.

In doing so, you enhance the user experience and foster deeper connections with your customers—customers who inherently want to buy from you, time and time again.

You possess the power to shape the type of experience your customers encounter on their digital journey. Whether you take advantage of that power to give your customers the experience they *really* need is entirely up to you.

WHAT'S NEXT

Optimization is not a one-and-done process. To implement the knowledge and tactics to the fullest, you'll need to shift your mindset to see optimization as an ongoing process.

Behind the Click explores the fundamentals of the psychological forces that guide customer decision-making. This book provides the why behind the reasons many different digital journey optimization and conversion rate optimization strategies work. While we do lay out many specific optimization strategies for the most important elements of your website, there is so much more to explore.

The more you work within and learn about digital journey optimization, the stronger your results will be.

If you need help and want to enhance your company's digital experience, visit thegood.com to learn more about how The Good can assist with optimizing your digital experience.

If you're managing this process yourself, we at The Good offer many tools to continue your digital journey optimization journey. Visit thegood.com/insights for resources to help you along the way, including hundreds of in-depth articles sharing strategies and case studies for optimizing the digital experience.

For a more comprehensive look into digital journey optimization strategies, consider diving into R. Jon MacDonald's other books:

- *Opting into Optimization: How Successful E-commerce Brands Convert More Customers, Increase Profits, and Create Raving Fans.* Explore a collection of principles that, when applied in a disciplined manner, have proven

CONCLUSION

to help e-commerce leaders capitalize on unprecedented market demand and build sustainable, thriving businesses. *Purchase your copies on Amazon or thegood.com/oito*

- *Stop Marketing, Start Selling: Your Guide to Doubling Online Leads, Customers, and Revenue.* Nobody watches TV for the commercials or visits your website to check out your latest marketing campaigns. If they're on your site, your marketing worked. Now, it's time to help them get what they came for. *Purchase your copies on Amazon or thegood.com/smss.*

If you'd like to have me speak at your next event, interview me for your podcast or article, or just say hello, you can reach me at *jonmacdonald.com*.

Thanks for reading. Here's to a better internet for everyone.

ABOUT THE AUTHOR

Jon MacDonald is founder of The Good, a digital experience optimization firm that has achieved results for some of the largest online brands including Adobe, Nike, Xerox, The Economist, and more. Author of three books on digital journey optimization and a frequent keynote speaker, he has been invited to share his expertise on important stages including Google and Autodesk. He knows how to get website visitors to take action.

As President of The Good, Jon helped lead the advisory to become one of Oregon's top 20 fastest growing private companies three years in a row. Jon is also a Forty Under 40 Award Winner, recognizing forty professionals under age 40 who have excelled in their field, shown tremendous leadership and are committed to their community.

Jon started his career as the sole web designer/developer for a start-up advertising agency during the dot-com boom.

Throughout his career he has engaged with dozens of brands including Autodesk, Apple, Columbia Sportswear, Comcast, Cisco, General Electric, Harley-Davidson, HP, Intel, Microsoft, Nationwide Insurance, Nike, Nokia, Red Bull, UPS, Vodafone and Xerox.

Jon volunteers for several causes throughout the Pacific Northwest and is an active committee member of industry associations and peer groups such as Entrepreneurs' Organization (EO). Jon is based in Portland, Oregon.

Printed in the USA
CPSIA information can be obtained
at www.ICGtesting.com
LVHW042033220524
780287LV00002B/2